Animal Behavior
AND ITS APPLICATIONS

by Derek V. Ellis

Library of Congress Cataloging in Publication Data
Main entry under title:

Ellis, Derek V.
Animal behavior

Bibliography: p.
Includes index.
1. Animal behavior. I. Title.
QL751.E498 1985 591.5'1 85-4279
ISBN 0-87371-020-7

Second Printing 1986

LEWIS PUBLISHERS, INC.
121 South Main Street, Chelsea, Michigan 48118

PRINTED IN THE UNITED STATES OF AMERICA

Cover photographs: (Top) Chacma baboons ranging over a surf beach. (Left) Migrating pink salmon nosing against the rear wall of a fishway pool. (Bottom left) Grouped sea slugs *Melibe leonina* feeding in an eelgrass bed. (Bottom right) Inuit hunters harnessing the energy of domesticated sledge dogs to home range and forage over arctic sea ice. (Photos by the author; original montage by Janice Friesen).

Derek V. Ellis

Derek V. Ellis is a Professor of Biology at the University of Victoria in Victoria, British Columbia, Canada. He trained as a zoologist at the Universities of Edinburgh, Copenhagen, and McGill, with graduate fieldwork in the Canadian arctic and Greenland. He started professional work as a researcher on salmon in 1957 at the Federal Government Pacific Biological Station in Nanaimo, B.C., becoming the first Canadian fisheries scientist to use scuba diving as an observational technique. He joined the faculty of the University of Victoria in 1964. He has taught animal behavior, marine biology and environmental impact assessment at the university since then, while involving himself in research on marine benthos and salmon, with digressions to wildlife, domesticated animals and primates.

Since 1970 Dr. Ellis has been involved in environmental impact assessments on the waste discharges of cities, pulp mills and mines. His graduate students presently research recolonization processes of impacted marine ecosystems. He has acted as consultant and environmental auditor to government agencies and industrial plants. He is an editorial writer for *Marine Pollution Bulletin,* and is editor/co-author of *Pacific Salmon: Management for People* and *Marine Tailings Disposal.* His book, *Subordinate Sex,* compares ranking behavior in animals and humans.

PREFACE

This book is about the daily activities of animals, i.e. their behaviors, and how we as humans impact and manipulate those activities. It is written for biologists, engineers, veterinarians, animal care personnel, park naturalists, biological administrators, and others who are interested in the applications of animal behavior.

The basic premise in behavior applications is that the scientist must have some personal familiarity with the body language of the animals. They can tell us a great deal, if we will look and listen carefully and systematically. I hope this book will provide guidelines to users in how and what to observe, and the significance of what they see and hear.

Derek V. Ellis

ACKNOWLEDGMENTS

I am greatly indebted to many people for assistance and comments during the preparation of this book, and to many more during the years it was unconsciously coming together. Specifically, I thank the following: Gretchen Moyer for critically managing production of the manuscript, including efficient word processing; Cami Faulder for library retrievals; students in my Animal Behavior courses at the University of Victoria; students and faculty participants in the University of Victoria Neural Lunches and Malacology Discussion Group; George Pasek for comments on the "Experimenter" effect; Kees Groot and other colleagues at the Pacific Biological Station, Nanaimo, for cooperation over 25 years on salmon topics; and Elaine Morgan and George Branch for discussions of the Aquatic Ape and tropical lagoons concepts. I am also very grateful to Tom Gore, photographer, of the Biology Department, University of Victoria, for photographic printing, to Chris Bowen and Elizabeth Gilbert for drafting many of the figures, to Al Denbigh for drafting the salmon identification figures of Appendix 2, Rick Wagstaffe for computer indexing, and to John McInerney, Chairman of the Department, for his continuing support.

CONTENTS

CHAPTER 1. THE BEHAVING ANIMAL: BASIC WORKING CONCEPTS.

An animal goes about its daily business just as you and I do. It finds food, escapes becoming food for another animal, ranges its habitat, removes skin disturbances and body wastes, sleeps, socializes and mates. This is its behavior.

Like physiology, behavior is an activity which can be investigated by science. We can document the actual processes of behavior by objective measures and descriptions. We can generate hypotheses on the physiological and evolutionary causes for the behaviors, and (at least for the physiological causes) test those hypotheses by appropriate experimentation. We can go beyond the natural history into understanding the how and why of behavior. We can also manipulate the behavior of animals to our and their benefit, or we can mismanage it to their and our detriment.

There are problems, however, in the scientific investigation of behavior whether for understanding fundamentals or applications. These problems are different from and more difficult to resolve than those arising from physiology. One of them is that much animal behavior is so similar in pattern or function to human behavior that we borrow everyday words to describe the actions. Many of our words, however, such as courtship, aggression, altruism and fright, have emotional implications which make them ambiguous to us. We must use descriptive terms which reduce subjectivity and ambiguity as much as possible.

Another problem is that behavior is the product of animal anatomy and physiology integrated in ways that allow the animal to survive within its ecosystem. Progressive investigation into behaviors eventually leads to the investigation of anatomy and physiology. Accordingly, investigators have to decide how far they need to involve themselves in the ecological, anatomical, and particularly physiological underpinnings of the actions. The problem is resolved by black boxing compartments of the behavioral model which the investigator implicitly or explicitly develops. Thus the ecologically-minded biologist whose interests lie largely in the field investigation of the survival value of behaviors, may decide to black box his or her understanding of the physiological mechanisms undoubtedly occurring.

Black boxing is an inevitable working concept, as it acknowledges phenomena which exist but which cannot be investigated at the moment. At its best, black boxing is a useful shorthand and reminder, although at its worst it is an unconscious assignment of irrelevancy. One of the best known early models of physiological concepts underlying overt behavior was developed by Konrad Lorenz, and has come to be called the flush toilet model (Figure 1.1a). It was controversial because it attempted to make a hydromechanical representation within the linked compartments of the nervous system (Lorenz 1950: 256). A black box form of the flush toilet model can be developed (Figure 1.1b). The model now indicates a sequence for physiological processes which need investigating to test whether Lorenz's hypotheses of action specific energy, releasing mechanisms, and other concepts, are useful in investigating behavior physiology. Some still are; others are not.

Black boxing is an important working concept for behaviorists, especially those interested in specific applications. We often use the approach and should explicitly acknowledge it as a means of graphical shorthand.

What are they doing?

When you walk, snorkel, or scuba dive anywhere in an ecosystem you see animals going about their daily business, and much of what they do has an obvious function. It is often easy to tell when an animal is feeding (it bites off and swallows), escaping (it darts away), or mating (it mounts or is mounted). The behaviors have obvious survival value in keeping the animal alive, and allowing it to perpetuate itself. They are performed in ways which are obviously adaptive to the ecosystem in which the animal finds itself. We feel that we understand what the animals are doing and why. We recognize that the animal is individually fit enough to survive.

But for some of the behaviors that we see, such as the self-sacrificing hostility of guard ants, it is not at all easy to decide what the survival value is for the dying individual. It is not easy for us to understand the behavior's adaptiveness, and certainly not its causation in terms of natural selection and evolution. Also, there are many obvious activities, such as the stereotyped flamboyant behavior of birds, which have only become widely recognized as functional - in courtship - since they were studied in the early 20th century (e.g. Pycroft 1913 and Huxley 1914). There are many other less obvious activities (which we can see, hear and even smell if we take the trouble to observe carefully) the causation of which and even function was or remains equally obscure. The complex disturbance responses of termites (Marais 1971), the echo scanning of bats and humans (Griffin 1958), sentry duty by a few individuals of social animals, leading-to-milking by a dominant cow, and pheromone release by insects, were not at all obvious until someone noted the unexpected. What is more, scientists often lag behind naturalists, fishermen, and farmers in their observations.

Many experienced behaviorists maintain that they still see novel behaviors when they observe their favorite species at well-known sites, or repeatedly view a film or videotape. For the novice behaviorist there is an enormous array of unexpected, and apparently inexplicable, behaviors out there in the ecosystem waiting to be observed, hypothesized about, and eventually "understood". You can see for yourself, like many others before you.

Charles Darwin is best known for his formal expression of the theory of evolution by natural selection, i.e. the characters of a species can change over long periods of time through the differential survival of some individuals. These individuals are "fitter", i.e. better adapted than others. These fitter individuals fix inheritable qualities in the species' gene pool. Darwin formatted his theory from a mix of anatomical, ecological and behavioral observations. He is an outstanding example of the scientist who can use the inductive method - the collation of many items of related information, and the formulation of a theory from and about them. Thus over a fifty year period, from the start of the voyage of the Beagle in 1831 (Darwin 1839), to his book on behavior entitled *The Expression of the Emotions in Man and Animals* (Darwin 1872), to his death in 1882, he collated and theorized about the behavior of earthworms (Darwin 1881), domestic animals (Darwin 1868), and other species. An introduction to Darwin is given by Huxley and Kettlewell (1965). He was an early entrant into the field of behavior applications as he was into so many other fields.

Observation and inductive reasoning is the starting point for scientific investigation. Without personal observation of the range of behavior of which a species is capable, even if we use the species only as a medium to investigate specific behavior of interest, e.g. utilization of rats to understand learning (Barnett 1963), we have not gone back to basics. We can easily be misled about the function and scope of the behavior, and how to investigate it by deductive reasoning and appropriate experimentation. It is very easy in a well studied field to begin where deductive experimentalists (applying theories and testing them in specific cases) have left off. But to do so without

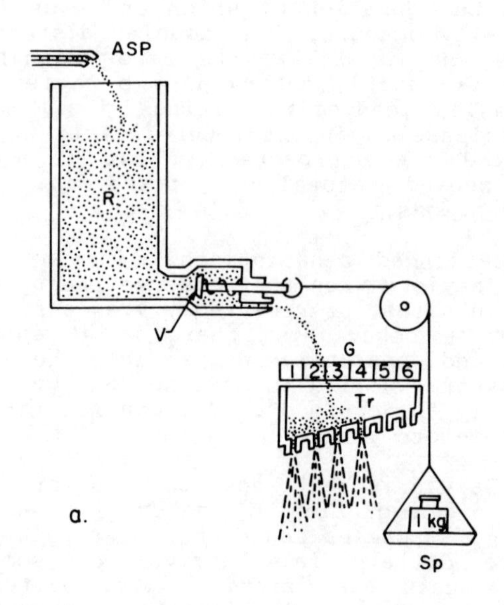
ASP
R
V
G
1 2 3 4 5 6
Tr
1 kg
Sp
a.

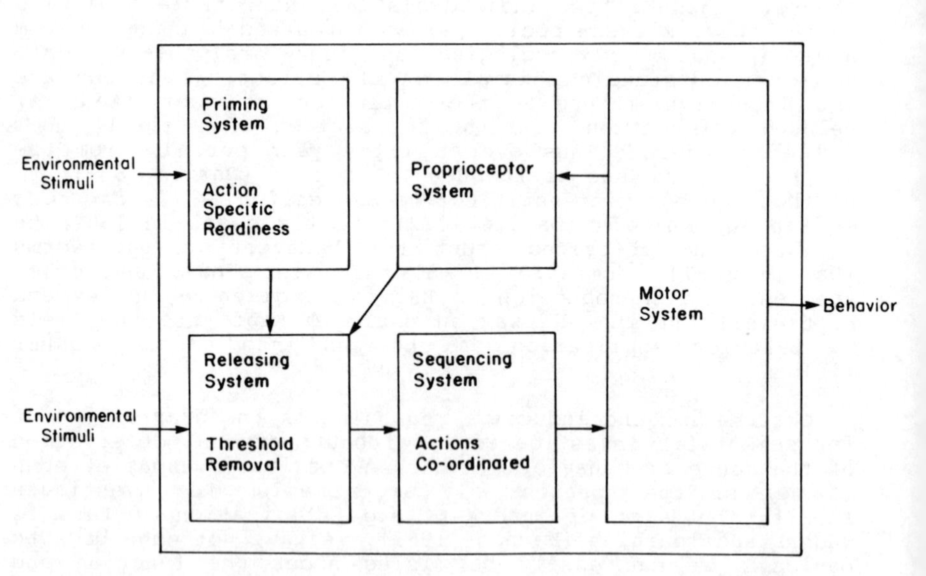
Environmental Stimuli
Priming System
Action Specific Readiness
Proprioceptor System
Motor System
Behavior
Environmental Stimuli
Releasing System
Threshold Removal
Sequencing System
Actions Co-ordinated
b.

Figure 1.1a. Lorenz's (1950) flush toilet model of stimulus response mechanisms involved in priming and releasing behavioral sequences. Action-specific energy (ASP) accumulates in a reservoir (R) and is released to become effective in producing a response when cone-valve (V) inertia (a threshold to action) is overcome by weights (stimulii) added to a scale-pan (Sp). Actions are sequenced by the accumulated weight (stimulus quantification) determining the level of cone-valve opening and where the ASP jet will fall onto a graded series of holes in a trough (Tr). G represents a scale measuring the response intensity. The scale-pan is located close to the trough so as to receive additional weight from the discharged liquid and provide a feedback information system.

Figure 1.1b. A black box model representing Lorenz's hypotheses. The essential compartments of priming, releasing, sequencing and information feedback are included, although concepts in the original model concerning sequencing have been eliminated since they require further compartmentalization within the sequencing box. The black box concept removes the hydro-mechanical graphics which in the original model caused so much argument.

observing basics for oneself often causes one to miss important clues for better understanding. Every student and user of animal behavior should observe for him or herself as much as possible.

Karl von Frisch conducted research in two very well-known fields: bee biology and sensory perception. He brought his results together in a book, *The Dancing Bees* (von Frisch 1953), in which it is obvious that he made enormous numbers of observations and simple experimental manipulations of bee activities over many years. He mentions in his preface, however, that forty-five years before he started his work a distinguished scientist (he had the grace not to name him) concluded definitively that bees were colorblind. Von Frisch realized that he could not believe both such a laboratory derived conclusion and simultaneously the theory of evolution since bees select only certain colors of flowers for nectar and pollen gathering. He embarked on his lifetime of study of bee behavior and physiology, which led him to demonstrate that plants pollinated by bees had flower colors within the ultra-violet to yellow spectral sensitivity of the bees, whereas others pollinated by nectar-competing butterflies and hummingbirds had true red flowers which lacked ultra-violet and hence were not visible in color to the bees. His field observations of colored flower preferences led to novel experiments to which the earlier laboratory bound scientist, lacking such an observational background, was not directed.

Another outstanding inductive reasoner in behavior investigations is Konrad Lorenz, whose studies from the 1930s (e.g. Lorenz 1937) reflect the strength of the approach when it is applied objectively. Konrad Lorenz is honored (he was a joint Nobel Prize winner with Karl von Frisch and Niko Tinbergen for medicine and physiology in 1973) as one of the founders of the type of approach to behavior studies which has come to be known as ethology. Essentially, the ethologist bases his investigations on observations of behavior in the field (or what is called an open-field simulation, e.g. aquaria for fish investigations). This leads to hypotheses which can then be tested deductively and experimentally either in the field or under controlled laboratory conditions.

The third of the three joint Nobel Prize winners, Niko Tinbergen, combined both approaches. Tinbergen formulated hypotheses about the function and physiological causation of such behaviors as courtship in different species (e.g. sticklebacks and sea gulls) using a combination of field and aquarium observations, which were interwoven with the experimental testing of his ideas. He used extractive methods (the removal of ecosystem components - habitat or companion) and intrusive methods (the addition of individuals or models, etc.) (Tinbergen 1951). His analysis of the sequencing of stickleback courtship (Figure 1.2) by quantification of inanimate models, stereotyped actions and the introduction of varying shapes and colors, is an example of the reductionist approach. He reduced a complex sequence of actions to component parts, each of which can then be described in detail. Their sequencing can be expressed by progressive counts per time. The model properties which release and direct them can be documented. All this leads in turn to improved descriptions and understanding of the whole process.

The strength of the experimental approach is that behaviors can be reduced to simple manipulated situations from which comes evidence about what sensitivities or activities are within the range of the animal's capability.

I.P. Pavlov brought to scientific attention the ways in which learning, or the modification of behavior through experience, could be investigated (e.g. Pavlov 1928). He developed surgical methods to allow healthy dogs to provide measurable responses, such as the amount of salivary juices secreted. From his experiments he developed the theory of conditioned reflexes. A conditioned reflex arises when previously indifferent ecosystem properties, through coincidence with stimulii producing a response, also come to produce the response. His early experiments undertaken with colleagues showed that the sight of food (i.e. conditioning to visual stimulii) could produce saliva, as could water colored to resemble a saliva-stimulating acid (Koshtoyants no date). Pavlov developed these portents of the conditioned reflex theory into controllable procedures with measurable responses. He also developed the behavioral concept that new neural connections of conditioned reflexes allowed an animal to build an optimal and flexible survival behavior for itself by coincidence of stimulii, where inborn reflexes were inadequate to provide responses for the complexity and variability of ecosystem demands.

Pavlov's initiative in learning studies has been followed by large numbers of scientists, most of whom work within the discipline of psychology, and whose studies illustrate a difficulty which must be faced by animal behaviorists. To what extent do demonstrated capabilities or demonstrated processes actually occur within the lifetime of a particular animal, or extend throughout the population of a particular species, so that they play a functional role in adaptation to ecosystem demands, and are actually drawn upon to maintain the animal alive. Learning capacity is available to many, if not all species, but may function only in particular situations where the trade-off between the survival value of many interacting unalterable and alterable responses comes down in favor of modifiability.

Within the area of learning theory, major advances in understanding and use by humans were brought about by the experimental approaches of B.F. Skinner, who demonstrated that the classical conditioning of Pavlov was a different phenomenon from operant conditioning (e.g. Skinner 1938), an experimental refinement of natural trial-and-error learning. Animals could be trained to perform totally unexpected and complex acts by reward reinforcement of incipient movements. Pigeons can be trained to push buggies or play table tennis! Skinner's working procedures are intuitively known to many animal trainers, and are the basis for our ability to domesticate, cage, and train animals, taming them or modifying their behavior in the wild, and even to train skilled gymnasts and athletes, whose complex acts are built by sequential training of component movements. Undoubtedly the potential for operant conditioning is present in many species, but whether it is drawn upon as trial-and-error learning needs species-specific, and possibly even individual-specific, investigation.

The phenomenon of sensory and behavior potential, as opposed to actual manifestation of sensitivity and activities when adaptive, is revealed in many experimental

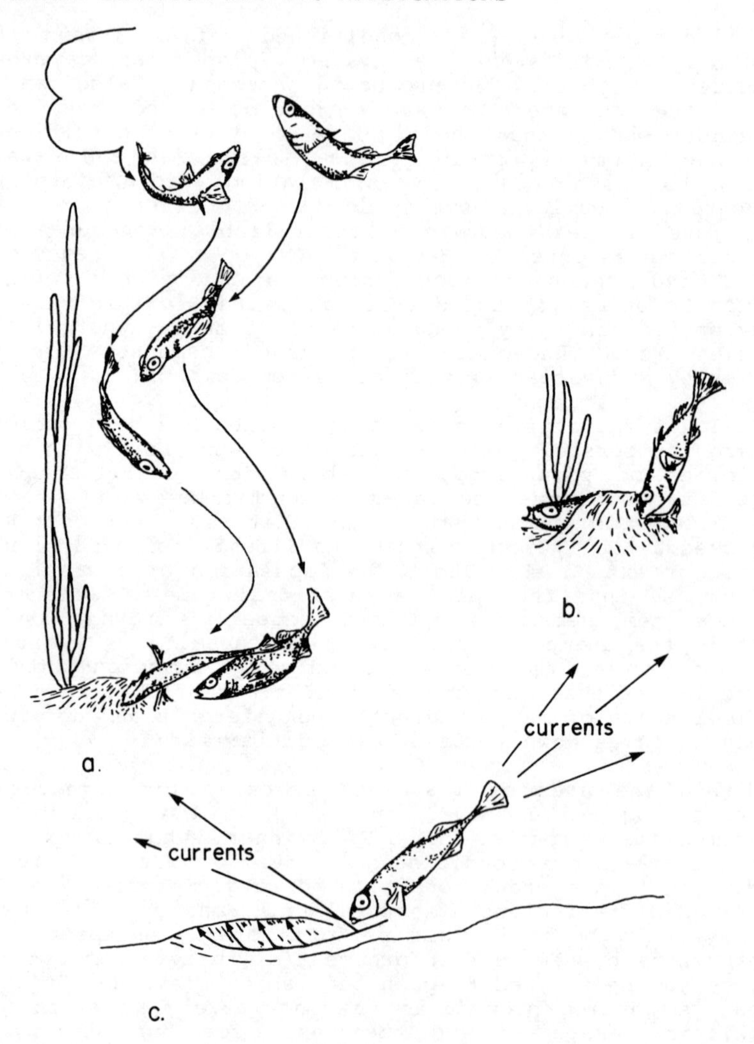

Figure 1.2. The male stickleback settles on and defends a territory while building and molding a nest of plant material. In the meantime a female locates him and unobtrusively waits in the plant canopy until the nest is completed. a. He then stimulates her to follow him to the nest by a behavioral display called the zig-zag dance. b. The female enters the nest and spawns while the male vibrates at the entrance. The behaviors of both sexes provide stimulii to the other sex, ensuring mutual sequencing of their actions.

The female leaves after spawning and departs from the territory; the male enters the nest, spawns and thereby fertilizes the eggs. c. The male guards the eggs, ventilates them, and subsequently guards the young as long as they remain together as a group. (Modified from Tinbergen 1951.)

investigations. The work of Bull (e.g. 1957) (Chapter 7) in drawing on conditioning experiments to determine the sensory capacity of fish, and of Fraenkel and Gunn (1941) (Chapter 8), which terminated in their review of orientation mechanisms (tropisms, taxes and kineses), both fall within the approach of demonstrating potential rather than realization.

Fraenkel and Gunn's review illustrates another important working concept for the behaviorist. Some books and papers are historically important because they provide a framework of definitions permeating subsequent investigations. If a later investigator ignores them he can confuse his readers if not himself, and even where subsequent studies eventually show the need for changing the concepts and terminology, the changes must be related to the previously accepted conventions. Fraenkel and Gunn's work, reviewed now, illustrates the nature of the change in the concept of taxes, and the manner in which the potential of animals to respond by kineses or taxes, must be considered in adaptive terms where other orientation mechanisms, such as configurational landmark recognition, are also within a species' potential. As an example, during the several months or years that Pacific salmon spend in large lakes or on the high seas, it is quite probable that they orient their overall directed movements with mixtures of temperature, light and chemical kineses and taxes, and configurational responses to celestial patterns or land forms. Each animal can draw on considerable species sensory potential, and whether it does so or not is a subject of field investigation separate from the laboratory sensitivity studies.

Since the late 1950s there has been a remarkable extension of balanced inductive and deductive behavioral studies on monkeys, apes, and to some extent, humans. These have been spearheaded by several female scientists who noticeably have undertaken many years' fieldwork in company with wild troops disturbed to the absolute minimum for observation only, without feeding, protection or with minimal experimental manipulation. The best known is Jane Goodall and her studies on the Gombe tribe of chimpanzees (e.g. Goodall 1971). From the starting point of prior concepts that primate social life was determined by the sexual drive of the male, she demonstrated the existence of substantially different and more complex group activities such as individual recognition, close kin bonds through matriarchal links, friendship relationships, group territories and warfare, tool use and learning, mild dominance and ranking, flexible group size, and incest prohibitions. Many other large mammals have now been subjected to similar long-term field observation in which the scientist concerned has come to know the individual animals, and has been accepted as just another non-disturbing animal within the group's ecosystem. They include such difficult species as elephants (Douglas-Hamilton and Douglas-Hamilton 1975) and killer whales (Ford and Ford 1981). The expanding list leads us to speculate what even more challenging species are now being or will eventually be subjected to the technique: polar bear, duck-billed platypus, bats, or perhaps eagles, crocodiles, sharks and sturgeon?

Another major recycling of behavioral interest came about from the inductive method during the 1970s. A number of theorists, collating their own and others' experimental, mathematical and observational results, developed a body of ideas now included under the general heading of sociobiology. The entomologist Wilson (1975) brought much of the material and ideas together. The working concepts stress objectivity in the interpretation of the adaptiveness and function of behaviors and their underlying structures, physiology, genetics, and particularly their evolutionary causation by natural selection. He reviewed and formalized developing ideas on altruism, that most refractory of behaviors to be understandable in terms of natural selection. If you, as a non-thinking animal, sacrifice yourself for another's benefit or even just reduce your chances of survival by actions (such as sentry duty) benefitting your group, how can such inherited behavior not be selected against, and hence disappear from the species gene pool? Much of the answer lies in inclusive fitness, kin selection and reciprocal altruism. The altruistic behavior benefits kin with similar gene type, or is reciprocated. The individual fitness of the animal may be reduced, but its inclusive fitness is increased, i.e. its actions contribute to the survival of kin bearing the same "altruistic" genes. Provided enough kin benefit, the genes are perpetuated. Wilson (1978) however, followed the trend set by several previous behaviorists and others (e.g. Morris 1967; Ardrey 1961); he applied his ideas to humans. This generated such considerable and ongoing criticism that "Sociobiology" as a modern term for behavior studies ironically has become too subjective for general adoption.

The units of behavior

"If you can't measure it, it isn't science". This is one attitude to science, but it presents a difficulty to behaviorists. How can you convert continuously sequenced activities largely brought about by locomotion, or at least the movement of body parts, into a set of parameters which can be measured, i.e. recorded quantitatively, in some way?

Fortunately, in many birds and fish and in some other complex animals such as mammals, arthropods and cephalopods, behaviors consist in part of actions which are stereotyped. This means that such actions follow a fixed sequence of muscular movements. Each time the action is performed it can be recognized as a unit of behavior. Thus the courtship displays of birds, fish and mammals fit into this concept, as do the threat gestures of species whose social life is mediated by dominance ranking. A male Pacific salmon responds to the redd digging of a female by vibrating alongside her. A baboon, gaping and exposing its fangs, threat displays to another in maintaining its position in the troop's rank order.

These stereotyped actions are called Fixed Action Patterns (FAPs). They are fixed in pattern to the extent of relatively little variation about a modal (most commonly repeated) action. Some are also fixed in timing. According to analysis from cine-film (Dane et al. 1959) goldeneyes (*Bucephala clangula*) perform a number of FAPs, many of which have a low standard deviation (Table 1.1) relative to the mean duration. Some have short and long forms due to variation in one or more components. Most are very quick, some taking less than a second to complete.

The existence of FAPs allows a number of measurement techniques. First and most important, FAPs can be recognized as unit patterns by any observer who has the patience to watch, and the support of film or videotape time-motion analysis. Each FAP can be provided with a name for convenience in recording and communicating. Each FAP can then be measured by its frequency of occurrence (no./unit time), duration (time), and in some cases by its intensity (angle achieved, number of dips, height of stretch, etc.). The FAP allows parameterizing behaviors without introducing artificial mechanisms such as levers to be pressed.

The existence of the FAP has some important physiological implications. A repeatable identical action, the separate muscular contractions of which are performed in sequence, implies physiological mechanisms releasing the action as a package, i.e. there can be an inbuilt system automatically carrying the FAP through to completion once it is triggered.

Table 1.1. Constancy of display actions of the goldeneye, *Bucephala clangula* as analyzed from cine-film (collated from Dane et al. 1959).

Display (sample size)	Duration in seconds	Standard deviation	Range in seconds	Comment
Usual flock displays				
Male				
Masthead(13)	4.80	2.50	2.90-9.40	Head-down variable, rest constant
Bowsprit(64)	1.75	0.30	1.25-2.55	
Nodding(14)	2.70	1.95	0.50-7.40	Number of nods varies
Simple head-throw(66)	1.29	0.08	1.08-1.50	
Head-throw-bowsprit(18)	3.00 or 1.94	? 0.22	2.95-3.00 1.70-2.35	Long and short forms
Head-throw-kick(49)	1.20 or 2.15	0.10 0.10	1.09-1.34 1.96-2.33	Long and short forms
Ticking(16)	0.19	0.05	0.13-0.28	
Head-flicking(21)	0.20	0.03	0.16-0.29	Terminates other actions
Female				
Head-up(17)	-	-	1.00-4min	Very variable
Bowsprit	-	-	-	Almost identical to male
Nodding	-	-	-	Similar to male
Head-forward(8)	1.22	0.33	0.85-1.80	
Dip(39)	1.79	0.75	0.40-3.54	

Table 1.1. continued.

Display (sample size)	Duration in seconds	Standard deviation	Range in seconds	Comment
Pre-copulatory actions and copulation				
Male				
Head-flick(11)	0.23	0.05	0.16-0.29	Occurs independently of other actions
Head-rubbing(18)	0.75	0.08	0.63-0.87	
Display-drinking(31)	2.02	0.11	1.90-2.25	
Wing-stretching(14)	4.80	0.90	3.75-5.90	
Bill-shaking(27)	0.70	0.09	0.60-0.95	
Display-preening(4)	0.60	-	-	
Pre-copulatory steaming(4)	-	-	1-3	
Copulation(3)	6-45	-	-	
Rotations(3)	3-6	-	-	
Post-copulatory steaming(3)	8-23	-	-	
Post-copulatory bathing	-	-	-	
Female				
Prone(9)	Several minutes	-	-	

Incomplete data means either highly variable or 4 or less observations or both.

It should not be necessary for each component part to receive further triggering from the coordinating center, but the action can proceed in an automatic manner. The example which has been quoted repeatedly is that of the greylag goose retrieving an egg displaced from its nest (Tinbergen 1951). The retrieval action of neck extension, and egg rolling by the bill, goes to completion even if the egg rolls away from the retrieving bill, so that in this case the FAP must be repeated until the egg is brought back to the nest where the next sequence takes over, ensuring that the egg is placed back under the body.

Much behavior of vertebrates has a train of FAPs put together in statistically variable sequences. Thus a set of courtship FAPs may be sequenced together in ways which are highly variable. Figure 1.3 shows the result of a sequence analysis of FAPs, and such sequences have been demonstrated for many species in many contexts. The general concept here is that complex, lengthy but synchronized social behaviors show variable sequences even when consisting of FAPs. The FAPs are invariable short-term packages of behavior, but the sequences are longer term packages during which there remains scope for variability to meet the demands of the particular occasion. For example, animals entering courtship may find their potential mates at different levels of sexual maturity and readiness to respond, so that the two have to be synchronized together. Predatory animals must modify their hunting sequence according to the nature of the prey on each occasion and the habitat they are both occupying at the time. Whatever behavior is undertaken, if it extends invariably over more than a few seconds, the time required for completion would render the individual less able to respond to dangerous situations. Thus variable sequences allowing interruptions permit escape from danger, and the opportunity to return to the sequence when the danger is passed. Sequencing, with variability, is one way of permitting the synchronizing of such potentially time-consuming social behaviors as mating and dominance ranking, without interfering with the survival imperatives of continual alertness against predators, feeding, drinking and habitat changes when necessary. Dane et al. (1959) and Table 1.1 show that some of the apparent FAPs of the goldeneye either represent short sequences of actions which can occur separately, or continue for so long that they are probably interruptable. The FAP and the long sequence of FAPs must be extremes of a range of physiological mechanisms releasing actions as an invariable package or allowing variable and interruptable behaviors.

Much animal behavior, particularly in non-social animals, does not include FAPs, and is far more difficult to parameterize and measure. A common approach available to experimental scientists is to introduce a recording device into the experimental situation. The best of these, such as a daytime light beam and photocell which can trigger interruptions in an electrical signal, are invisible to the animal under experiment, but considerable ingenuity is needed to ensure that the artifact measures a parameter relevant to the hypothesis under test. For instance, a measure of the frequency in which a fish interrupts a set of light beams directed along the bottom of its aquarium, will not measure its activity near the surface. Another kind of introduced device is the lever which must be pressed, or electronic pad which must be touched. These devices commonly rely on a preliminary conditioning procedure training the animal to trigger the recording system. An alternative approach is to cage the animal in ways which force stereotyped behaviors or routes upon it, and record frequency, etc. of such directed behaviors.

In each case a question which should be answered is: what else is happening that the recording system is not measuring? There really is no substitute for observations of the total behavior of the animal in the experimental situation to provide further ideas about the behavior which has been constrained into the observer's working equipment by his working concepts.

The most difficult behaviors to measure and describe objectively are those of non-social animals, particularly soft-bodied forms without skeletons to constrain muscular contraction into set leverage patterns. The swimming behavior of a jellyfish, the crawling of a flatworm, but also the random swimming of a non-feeding open water fish can be difficult to parameterize in unarguable ways. However the need is there, and the investigator must respond by tentatively assigning unit status to such actions as "bell-pulsing", "slow crawling", and "random cruising", supplemented by cine-film or videotape analysis if necessary. It must be recognized that these units are not FAPs in the traditional sense since there is so much variation in the orientation or duration of the actions. They are best "understood" (black boxed) as behaviors which can be released and will perpetuate themselves by some physiological system unless otherwise interrupted.

The non-social behavior of lethargic domestic animals bred for meat production, and other social animals, should be considered in the same way since they do not readily fit into FAPs. Such actions as "walking" of cattle, "pecking" of fowl, and "home-ranging" of dogs are definable and measurable (in terms of duration and frequency) but interruptable at any time by other actions.

While developing a system of measurable behavioral units for any species, the investigator needs to name each unit for recording purposes. This also presents problems for which there are some working principles. The apparent ideal

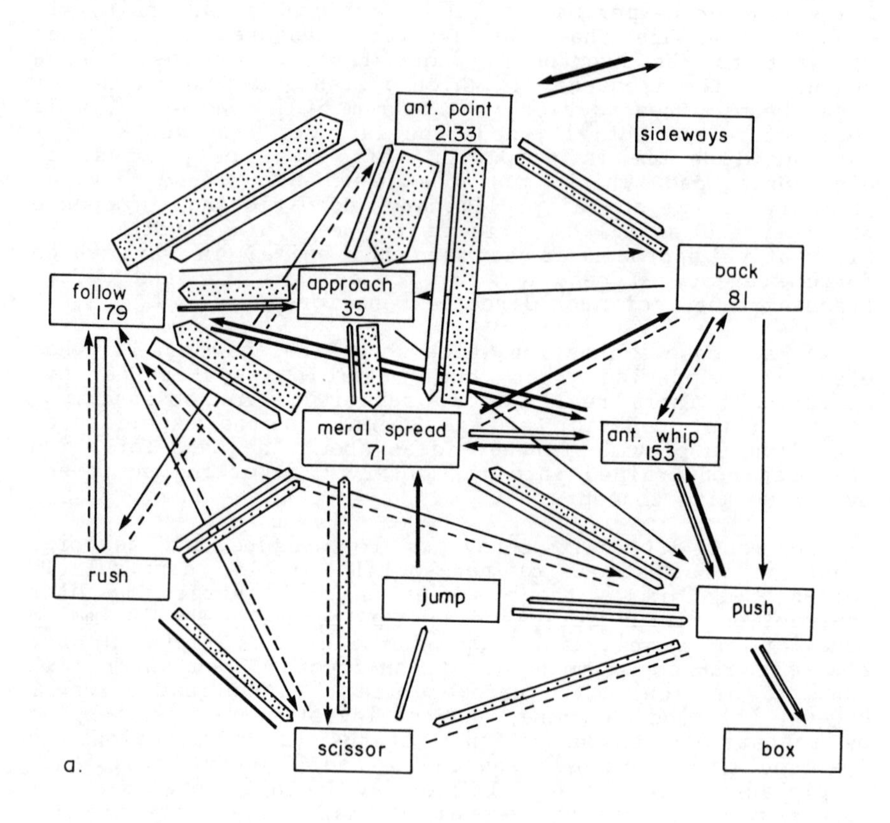

Figure 1.3. Sequence analysis of FAPs among lobsters in aggressive encounters. Numbers are frequencies (values less than 30 are omitted), and bar width is in proportion. Data and diagrams are from an M.Sc. thesis by C. Scrivener (1970). A represents sequences followed only by winners, and B sequences followed only by losers, i.e. individuals that retreated or avoided the other. The predominant sequences for winners and losers are clearly different, and give predictive capability based on whether the response to initial "antennae point" is "approach" or "back".

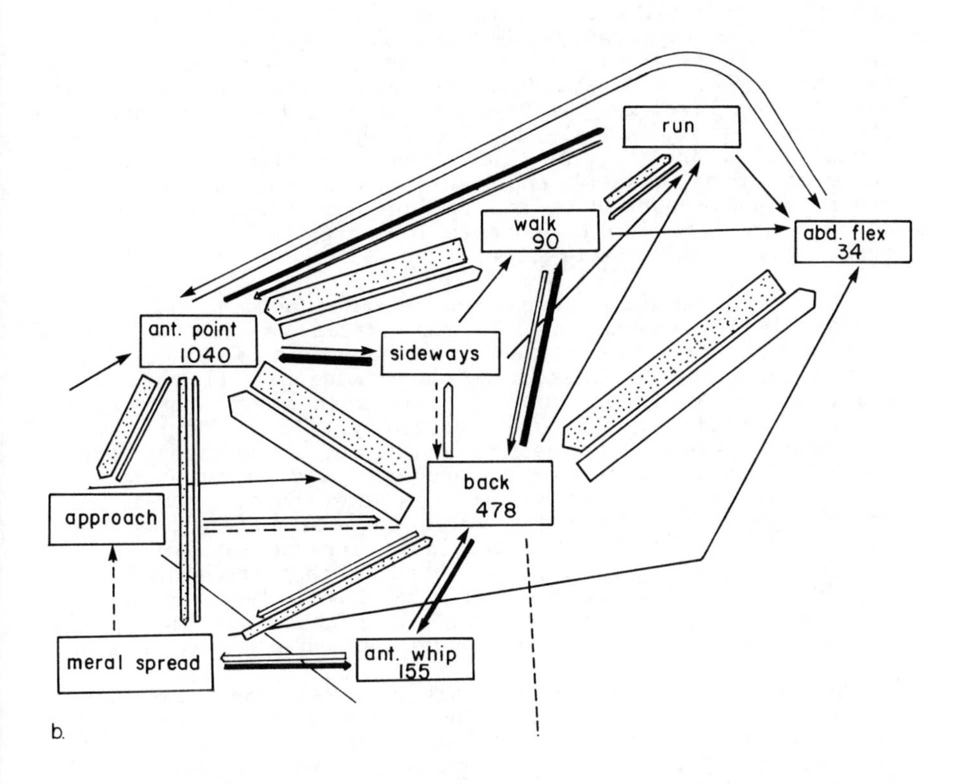

of numbering the units for objectivity is too hard to follow and too boring for many behaviorists, or at least for their colleagues to whom they will be presenting their results. Names have been applied which have emotional, subjective overtones, and if these are too pronounced they can be very misleading in directing the investigation of the causes of the behavior. Thus the term courtship behavior is at the limit of objectivity in what is acceptable naming. The term is now commonly taken simply to categorize actions leading to mating. We have thereby avoided the emotional human implications of mutual love between the courting individuals. Such terms as "the love dance" which appear in TV animal spectaculars are beyond the acceptable limit of descriptive naming. In sexual and power activities, terms such as "cuckold", "homosexual" and "runt" which are acceptable to our male dominated community of behaviorists, can be functionally and emotively misleading.

Descriptive terminology is the aim in naming behavioral units. The names should describe as objectively as possible the actions involved. Thus terms such as "stand-up-wing-stretch" are impeccably descriptive. They are a form of shorthand in that they simplify the total action. Emotive terminology is to be avoided. The distinction is relatively easy at the level of the FAP, but more difficult to draw at the level of functional sequences, where we draw on common human behavioral terms such as "feeding" (with its connotation of hunger) and "fright responses". Here the observer and reader have to realize that the term need not carry its emotional connotations with it. It is descriptive, not interpretive.

When parameterizing behaviors, we may need to express a hierarchical arrangement. In many cases certain behaviors only occur when the animal is engaged in a longer term course of action. Thus only when feeding will an animal seek, chase, kill, and devour, only when migrating will it "cruise steadily", only when escaping will it "dart away" and then "hide". In each case there is an upper level behavior (feeding, migrating and escaping) encompassing a set of lower level behaviors nested within. There are obvious physiological ideas which have been developed from the concept of the hierarchical arrangement of overt actions. The general sense of these is that the animal must be physiologically primed for the lower level behaviors which otherwise will not occur. Once these are initiated, the behaviors can be included under the upper level term.

The work of Packard and Ribic (1982) on sea otters provides an example of what can be achieved in the way of objective parameterization of behaviors, so that the form and variability of long sequences can be recognized, documented and communicated, thereby leading to the development of hypotheses about function and causation. These scientists grouped action patterns by form, not function, into activity states, thus recognizing a hierarchy of behaviors. The action patterns such as "somersault", "nosing", and "periscope" are more or less equivalent to traditional FAPs and the equivalent less stereotyped lower level actions mentioned in the last paragraph. The activity states "feed", "rest", and "groom" are equivalent to the upper level actions. Packard and Ribic were able to show by multivariate statistical methods that some action patterns only appeared in certain activity states (= clusters of action patterns), whereas other action patterns could appear in several activity states. Once again, parameterization, in this case with frequency measures and statistical analyses, revealed patterns of action from which the authors hope progress can be made in understanding their functions.

Parameterization of behavior raises the need to understand statistical concepts of accuracy and precision in a behavioral context. Accuracy is the concept of observing and describing the details of what the animal was actually doing. The two errors of describing behavior which the animal was not actually performing (Type I), and missing behavior which it was (Type II), are avoided. Precision is the ability to describe variability in the behavior. Thus Dane et al. (1959) appeared to have precise measures of goldeneye display durations due to the low standard deviations of the duration measures (Table 1.1). Their descriptions also appeared to be accurate according to the detail provided and its agreement with others' observations. There is yet another facet of accuracy and precision in behavior studies and that is the "experimenter effect". The experimenter (or observer) can affect the behavior of his subject animals by his presence and his equipment, or he or she can be influenced in their observations by experience (or lack of it), preconceptions and instrumentation available. Considerable care is needed to reduce errors of these various sorts.

Understanding behavior

There are times when we like to think that we understand a behavior. We mean that we have a hypothesis (which we may have tested, not disproved, and thus accepted) about its survival value to the animal (its function). Or our hypothesis may be about the physiological mechanisms which brought about the behavior (proximal causation), how it evolved (ultimate causation, or phylogeny), or how it developed (ontogeny). These four components of understanding are not clearly separate. The survival value of a behavior is the extant product of an evolutionary process which has shaped the function by continual selection pressure. Proximal causes encompass levels of understanding ranging from neurophysiological concepts arising from electronic probing to hormonal biochemistry. Developmental causes include DNA determination mechanisms for phenotypes to maturation and learning processes. Nevertheless, these four components are all concepts of understanding behavior.

The function of a behavior encompasses several concepts. It may mean simply the value of the behavior in keeping the animal alive, allowing it to grow and sexually mature, or in perpetuating itself. Some behaviors have obvious survival value. Feeding, escape, and mating are so obviously functional, with survival value, that the behavior concerned is named in a way which summarizes its function. "Incipient bite", "motionless", and "vibrating" are behaviors whose functions are not obvious from their names, and may not even be obvious when observed in their normal context. Nevertheless, it is inherent in our overall concept of natural selection that such behaviors will serve a function of some sort. They will be adaptive to the demands of the animal's ecosystem. Otherwise, they would have disappeared either by elimination from the species' gene pool or by temporarily not being expressed (but still in the gene pool) through some genetic/ontogenetic mechanism.

The function of a behavior cannot be considered in isolation from its value to the animal in keeping it alive and perpetuating itself, which in turn means its adaptiveness, and how it contributes to the "fitness" of the animal. The latter two concepts are not simple. They represent a balance of advantages and disadvantages to the animal, which resembles in some ways a benefit-cost relationship as found in the economic actions of humans. If the behavior is to be maintained in the animal's and hence the species' repertoire, then the balance between survival benefits and risk or energy costs must favor survival. Thus the courtship display actions of territorial birds and fish commonly flash species-specific or dominant male patterns, and these temporarily reduce cryptic coloration, on which the animal depends for its survival against predators. The duration and frequency of flashing such courtship patterns must balance the need for synchronizing mating actions with a partner of the same species against the need for remaining hidden from predators. The balance is conceptualized in Figure 1.4. The trade-off may take many forms, depending on whether, for example, one-shot mating transfers sufficient sperm to ensure production of the next generation or whether repeat matings are needed.

The benefit-cost balance of a behavior is also determined by its interrelationships with other behaviors. This is why it can be so difficult to show the adaptiveness of a discrete behavior. Nevertheless, it is so entertaining to speculate about behavioral functions and adaptiveness that we do it time and time again. Thus baldness in humans could be a sex-age signal implying "this is an experienced male, follow him (and the highly visible bald spot) in dangerous situations". Such a behavior could have had strong survival value in human societies exposed to predators and intertribal warfare, thus being adaptive and contributing to the fitness of individuals who carried the signal and those who responded to it. But how can we test such a hypothesis when the conditions in which we hypothesize that it was adaptive no longer apply? We cannot test it in our present society, although anecdotes about baldness or the lack of it in present hunting and gathering societies can serve as the basis for people to accept or reject the idea.

This concept of a benefit-cost balance needed between many behaviors is close to the concept of an Evolutionary Stable Strategy, i.e. a behavioral strategy which is stable from one generation to the next (Chapter 10). Under a constant environment any deviation from an ESS is less adaptive (Maynard Smith 1975, 1978). A behavioral strategy represents a complex of actions integrated to be adaptive

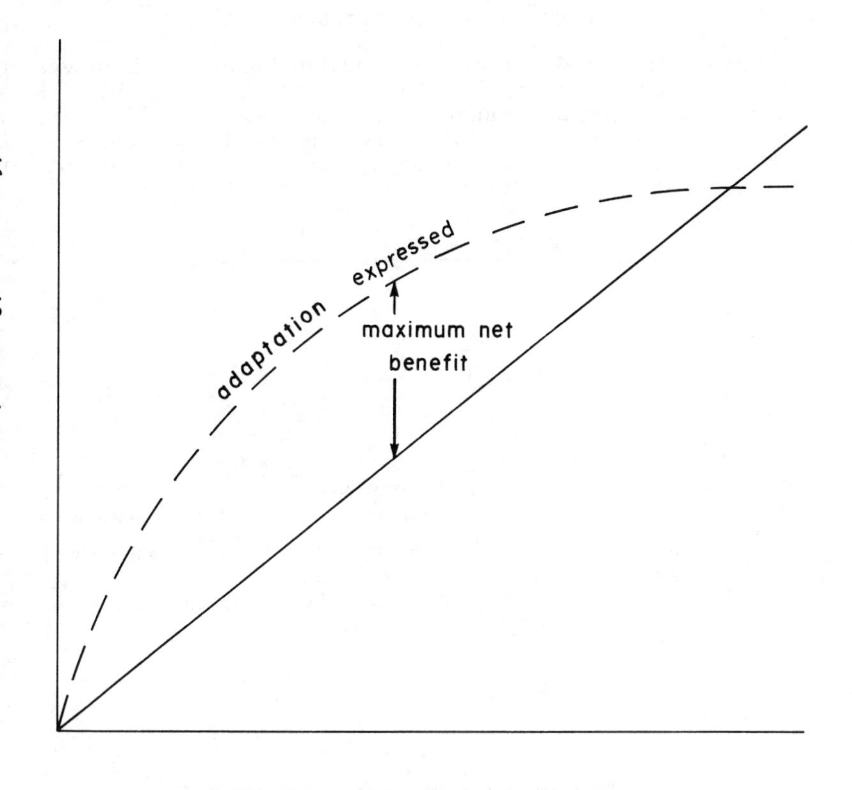

Figure 1.4. The benefit-cost concept applied to biological adaptations shows that any adaptation has costs in terms of energy expenditure and risks, etc. These are trade-offs against the benefits of survival and gene transmission to the next generation. Benefits may be measured by many different parameters -- age at death, mating frequency, fecundity, etc. The different currencies for benefits and costs limit the use of the analogy. Potentially there could be a point of maximum net benefit in the expression of an adaptation. Under evolutionary selection if the adaptation were exaggerated out of proportion to other interacting adaptations, there could be a point above which benefits were less than the costs.

under particular ecosystem conditions.

The next level of understanding behavior involves the physiological mechanisms which mediate it. This is the level of proximate causation. A model of the behaving animal is needed to black box the basic mechanisms, and indicate the areas within physiology that must be explored for understanding at this level.

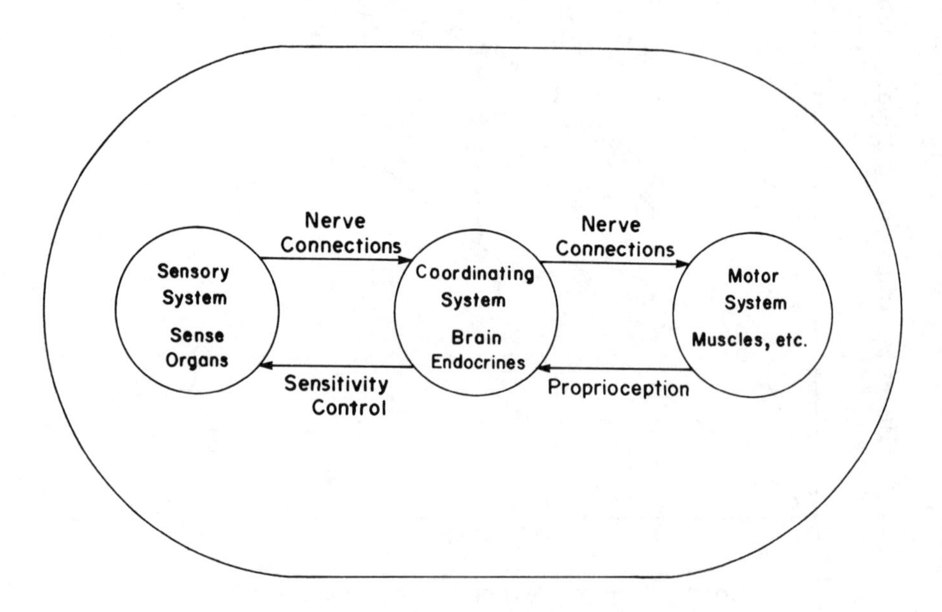

Figure 1.5. Black box model of a behaving animal. There are three black boxes (compartments) representing systems of sense organs, a coordinating center and motor organs, and linked by information transfer connections.

Figure 1.5 presents such a black box model. There are three major compartments - sense organs, a coordinating system and motor organs - linked by information transfer lines derived from the coordinating system.

The input consists of information from the ecosystem received by compartment 1: the sensory organs. The information can consist of signals from the inanimate environment, from conspecifics or other species--predators, symbionts, etc. The sense organs are located where functional, and in most active animals, e.g. vertebrates and insects, are assembled facing forward, as there is a need to sense the environment ahead. The sensitivity can be under the control of compartment 2, the coordination system.

The nervous system transfers information from the sense organs to the coordinating center (brain, cerebral ganglia, etc.) where it can be processed in many ways. The coordinating center may barely respond, in which case a more or less direct connection to the motor organs produces a reflex response. The coordinating center may process the received information, compare it with stored information (memory), and initiate complex behaviors. There appear to be filters within either the sense system or the coordinating system so that much of the enormous jumble of receivable stimulii perpetually being sensed is not responded to (thresholds and habituation). The sensed information may be transferred to a subcompartment of endocrine organs, which by flooding an alternate information transmission (vascular) system with chemicals, can mediate relatively slower and longer lasting responses than the nervous system.

Finally, information is passed to a set of motor organs producing a response, which hopefully is adaptive, and contributes to the animal's survival or breeding. A feedback channel by internal proprioceptive sensors allows modification, particularly by orientation fine-tuning responses to the stimulii. The feedback for termination may be external, as the animal receives from the ecosystem information on consequences of its released behavior. The motor response may be released as a package and go to completion (the FAP already described) or be subject to termination by neural control. The motor organs generally consist of muscles operating with a set of skeletal levers, but include cellular organelles such as flagellae and cilia. It is also impossible to draw a dividing line between physiological and behavioral processes involving such motor organs as secretory glands producing, for example, silk, which is then woven into useful structures by appropriate behavior, and luminescent organs which can flash on and off in patterns, functioning as stimulii in courtship and predation.

These latter cases illustrate the need for many behaviorists to work at the interface with physiology, and similarly for many neurophysiologists and endocrinologists to expand their understanding of the adaptiveness of internal chemical and physical processes by engaging in behavior studies. In contrast, many behaviorists are content to reduce their understanding of the physiological processes to the black box state, at least temporarily, while engaged in studies of function and ultimate causation. The decision where to limit one's understanding is personal, and changeable.

Thus we understand proximal causes when we have revealed the physiology and biochemistry by which the behavior is brought about. We understand how the behavior is mediated. There is also the kind of understanding called "ultimate causation". Why does the animal behave the way it does instead of in other equally probable or improbable ways? We need to understand why the animal does what it does. This is much more difficult than understanding how it does it.

In a sense the "why" question can be answered satisfactorily for religious people by referring to their God, but this of course ignores the question of what mechanisms their God uses. The mechanisms are evolution and natural selection, i.e. inheritable changes plus selection pressures determining that the fitter animal will contribute a disproportionately large part of the next generation's gene pool. Thus our understanding of the causes of behavior returns to the question of their function and adaptiveness, but now in this context the process of evolution must be understood (or at least outlined in black box terms).

Figure 1.6 models the way natural selection works over a few generations. The generation 1 (G1) ancestor in stage 1 lives and is adapted to demands for survival set by its ecosystem. The ancestral individual is well adapted and has successfully competed for optimal conditions which balance optimums for all the ecosystem's demands. As a result, it is more likely to perpetuate itself than a less well adapted animal living in marginal conditions, which is more likely to be subjected to limiting stresses through oscillations in ecosystem conditions during its lifetime. In the context of temperature as an ecosystem stress, animals in marginal conditions are more likely to suffer from an extremely cold winter than those in optimal conditions. G1's many offspring compete among themselves and with other members of their species, and only some manage to inhabit the optimal conditions like their parents. Others are forced to marginal conditions, and still others may run foul of limiting factors. These limits may directly kill. Animals do freeze to death occasionally, but more commonly the effect is statistical - a reduction in the probability of breeding through such effects as reduced fecundity through starvation, or a reduced opportunity to breed through the lack of finding a healthy mate. Some G2s may even have better resistance to cold than G1 so that they and their descendants can survive reasonably well in conditions which are marginal for the species as a whole.

These effects will repeat themselves generation after generation while the ecosystem demands remain constant, or at least remain within an equilibrium to which the animal and its descendants are well adapted.

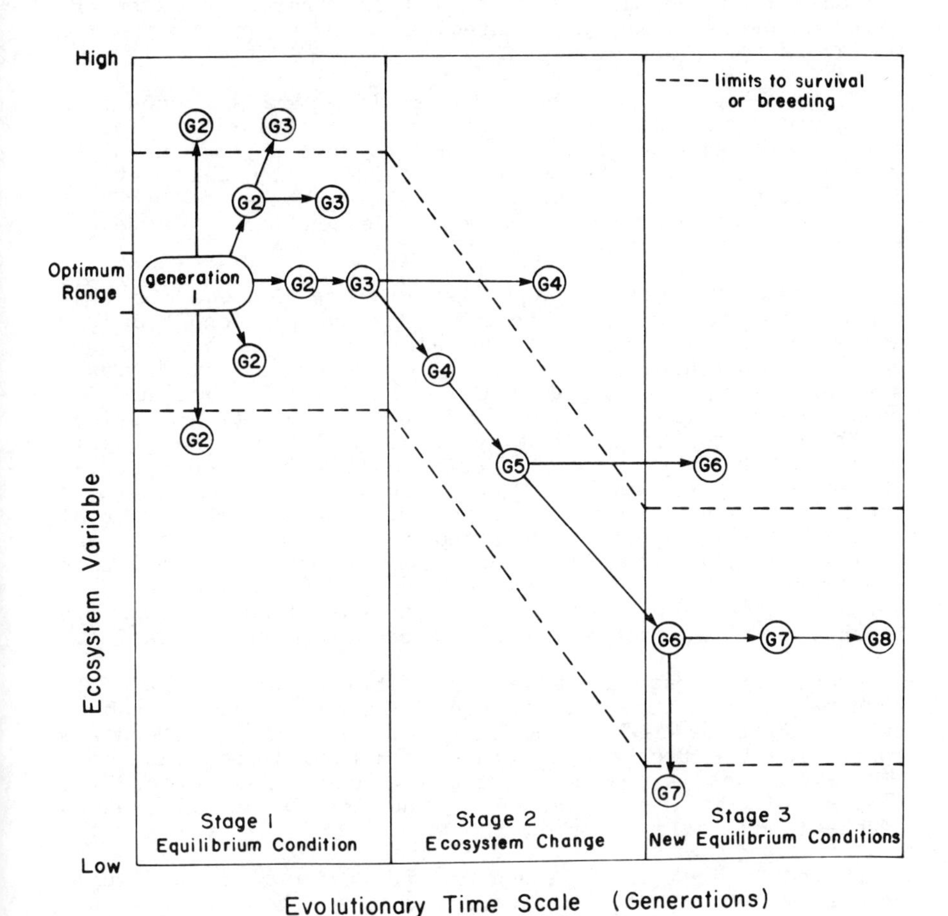

Figure 1.6. Model of evolutionary responses in behavior as succeeding generations from an ancestor adapt to changing ecosystem demands. A simple demand is correct environmental temperature. A change of evolutionary consequence is the cooling occurring during Pleistocene ice ages. (More detailed explanation in text.)

Ecosystems change, however, partly through physical and chemical changes in the environment, and partly by biological changes in the community as other species enter, exit or evolve counter strategies to the species as the predators, prey or symbionts.

Stage 2 in Figure 1.6 models a period of ecosystem change, such as when advancing glaciation lowers temperatures in the G1's home range to levels where even the warmest conditions of summer are colder than the previous lower limit of winter. During this stage there can be three behavioral consequences. The G1 descendants can desert their location, be eliminated or adapt.

The new ecosystem demands favor those descendants that express the potential to tolerate the colder conditions in the same way that some of the earlier G2 did and thereby came to inhabit marginal temperature conditions. Effectively the later generations shift their optimum for temperatures within the overall balance of ecosystem optima. If they can cope with this by one or other of the genetic processes which go on (gene recombination, mutation, etc.), then later generations (G4 and G5 in Figure 1.6) which previously would have been maladapted to the initial ecosystem demands are now adapted to the increased cold and will survive, unlike the G4 or the G6 which cannot maintain the trend. The process of changing ecosystem limits would normally take many generations if successful adaptations are to arise without the descendants abandoning their location.

Eventually Stage 3, a new equilibrium, arises and from then on further generations repeat the processes in Stage 1 but do so to a new system optimum. However, descendants continuing the trend to tolerate colder conditions are more likely to inhabit new marginal conditions and so exceed the limits of their tolerance. The trend to increased adaptation to cold is stopped.

While Figure 1.6 represents adaptation to a single ecosystem demand, that of temperature, and the changes that can occur in a location with an environmental change such as climate, the temperature adaptation has to be balanced with other consequences of such physical change (such as predators adapting more quickly), and with other coincident ecosystem changes unrelated to the temperature effect, e.g. the immigration of new predators due to land bridges. Evolutionary changes representing such an inevitable mix of adaptations to changing conditions are difficult to prove by the scientific method. Our understanding of evolutionary causes of behavior is essentially inductive in that we usually can only collate fossil records and their behavioral implications, and compare the behavior of taxonomically related species using the concepts of classical phylogeny. It is only rarely that an evolutionary process has been found to occur so rapidly that it can be experimentally analyzed. A well understood case is that of the peppered moth *Biston betularia* (Kettlewell 1959) and its cryptic darkening and behavior in response to decades of atmospheric pollution by a process called industrial melanism (Chapter 10). In addition, processes of artificial selection by breeding of domestic animals have been equated to natural selection processes with the changing limits to ecosystem demands under human control (Chapter 6). If we can breed to produce animal adaptations to our demands, we have at least some practical understanding of the processes involved.

These evolutionary processes may extend over the whole range of the species, significantly influencing the whole species' gene pool and its biology. However, the processes may influence only a part of the species' total population, in which case if additional mechanisms operate preventing breeding outside the stock showing the adaptation, a new subspecies and eventually species may evolve in addition to the ancestral form elsewhere. The speed of the process also appears to vary, giving rise to geological periods of greater or lesser species and ecosystem diversity. This topic is taken up in more detail in Chapter 10.

The concept of "inclusive fitness" may also operate in this sequence of events. The G2s in Figure 1.6 can establish in their kin group which of them will occupy optimal, which marginal and which intolerable conditions by damaging fighting or without harm by ritual contests. Peacefully dispersing reduces the "individual fitness" of the ones taking up suboptimal conditions, but not the fitness of the fittest. This process contributes to the "inclusive" fitness of the rest in the sense that the fittest of their kin, in competition with non-kin, have a high probability of passing on their shared inherited properties to yet another generation. The "inclusive fitness" of an animal is the sum of its individual fitness and the effect its actions have on the survival of kin carrying the same genes.

A further level of understanding comes through ontogeny: knowledge of the manner in which a behavior has developed. Figure 1.7 models the concept that behaviors in the adult breeding animal exist as a result of development both in the behaviors and in the structures and physiology on which they are based. Anytime the behaviors are expressed they must be functional, sufficiently enough so as not to detract from the fitness of the animal. Many behaviors, particularly breeding behaviors, are expressed functionally and perfectly the first time late in the life of the animal. The behavior itself does not develop in form, but is only shown in proper form when the structures and physiology are finally in place. There are two variations on this theme, as shown in the model. A behavior may develop in form, i.e. be learnt or mature through practice, as kittens and other young

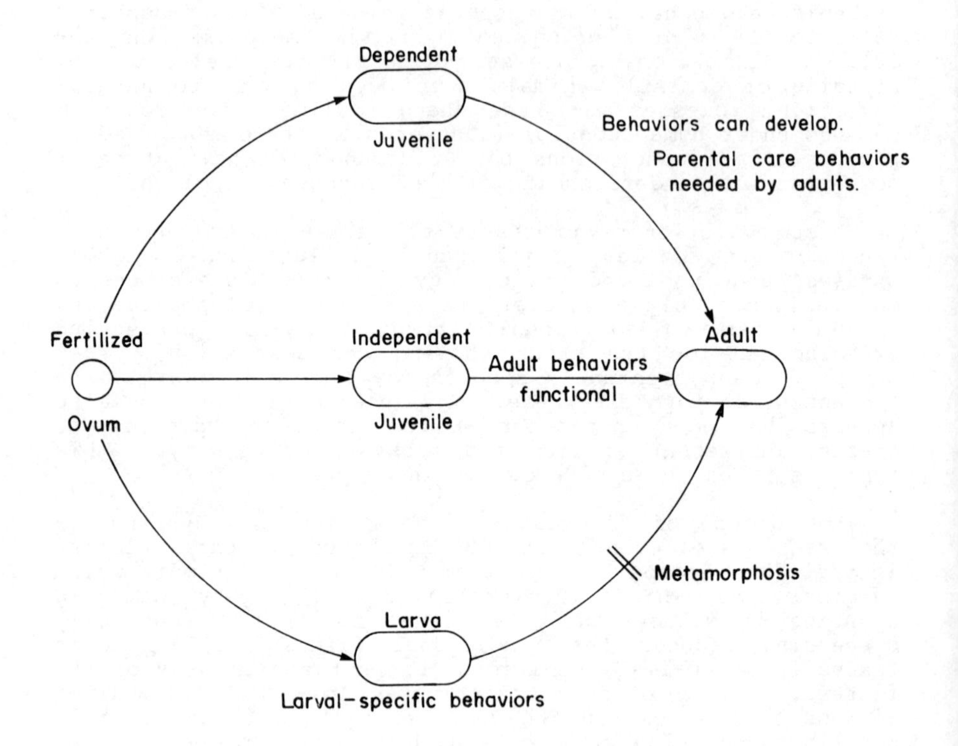

Figure 1.7. The functional behavior of the adult animal is achieved by three kinds of ontogeny. Independent juveniles (little adults) have essentially the same range of functional behaviors. In dependent juveniles the range of functional adult behaviors appears slowly and may be imperfect initially. There must be compensating factors in the ecosystem such as parental care or protected habitat. The third ontogenetic route to adult behavior consists of adaptation to a non-competing niche in the same or another ecosystem, with behaviors which are functional there. Change to the adult (breeding) stage may require a metamorphosis or dramatic change from larval to adult form and behavior.

carnivores learn or acquire the advanced skills of hunting. In these cases, there must be some feature of the ecosystem, such as parental care, which compensates for the ineffectiveness of the behavior in its early stages. The other variant is that some behaviors are functional only in certain life stages, and only occur there, not earlier or later in another stage. An example is the feeding of some caterpillars, which gives way to the non-feeding breeding stage of the butterfly or moth. When expressed the behaviors need to be functional, but as the animal changes and shifts its ecosystem, behaviors can be eliminated from or brought into the repertoire, often by a pronounced metamorphosis histolyzing, eliminating and reforming structures on which they were based.

There are three kinds of "understanding" of animal behavior which are either not very helpful or not experimentally verifiable. In practice, they often lead to complete fallacies which are downright wrong.

The first is anthropomorphism: "understanding" animal behavior in terms of human emotions and subjectivity. An example is the explanation that the dog "feels guilty" when it cringes before or avoids our gaze as we confront it with some damage. Human emotions are the product of human cerebral hemispheres, which are uniquely developed in us. The most we might be justified in hypothesizing is that there is some vague forerunner of emotions located in a more primitive part of the brain and associated with functional behavior, so that an animal might feel, in appropriate circumstances, stirrings of what we call fear, pleasure or guilt in ourselves. It is most unlikely that these are felt to the level where they have become Pavlovian conditioned reflexes and thus proximal causes for the behaviors. Certainly dogs and some other animals can show behavior similar to that which we perform in circumstances where we have strong emotions, but testable understanding lies in such functional explanations as: the dog has learnt (is operant conditioned) that punishment comes unless it shows cringing behavior and gaze avoidance. Male behaviorists also need to guard against unconscious chauvinist interpretation of dominance and sexual behavior (Gowaty 1982) in animal social behavior, e.g. female chimpanzees are promiscuous (i.e. enjoy it) rather than maximizing chances of being mated by a superior male.

The second error is anthropocentrism: "understanding" behavior as centered around the human good. "Dogs and cattle, fish and game are here on earth for human benefits". At its best the error of anthropocentrism merely masks the better understanding that animals can be manipulated individually by taming, etc., and in masses by domesticating and stock management. At its worst anthropocentrism leads to cruel exploitation, wastage, and species extinction, such as found in whale hunting, mindless pollution, and the virtual or actual loss of useful and vast gene pools such as the bison and the passenger pigeon.

Both these two errors include the third, that of teleology. Teleology is the fallacy of "understanding" consequences as causes. The "understanding" is circular and hides the need for better understanding by verifiable testing. Thus the statement that "an animal did so in order to" is inherently teleological in a proximal sense, but may not be in an ultimate sense. "Why did the chicken cross the road"? "To get to the other side" or "for some fowl reason" are both teleological statements, and the latter anthropomorphic. It crossed the road in terms of proximal causation because some stimulus over there attracted it, or some other stimulus on this side drove it across. In ultimate causation terms it crossed the road since crossing instead of not crossing increased the chances of obtaining food or escaping from a predator and hence had survival value. Where "in order to" or simply "to ....." occurs in a behavioral context, suspect teleology and the limited understanding of thinking that consequences are causes. Replace "to ....." with "and as a result ....." and start dissociating consequence from proximal cause and relating it to ultimate cause.

Another common example of teleological thinking is that an individual animal can act for the good of the species or the group. It doesn't. The benefits to the species or the group are a consequence of the action, not its cause (see Chapters 4 and 10).

Summary

Animals behave in functional ways in the sense usually of movements through their ecosystem. Their behavior is amenable to scientific investigation by measuring, recording, objective description, data collation, hypothesis generation and experimental testing. The two complementary methods of scientific investigation, inductive and deductive reasoning, must be applied, and were to varying degrees by the Nobel Prize winners Von Frisch, Lorenz and Tinbergen, and by others such as Darwin, Pavlov, Skinner, Goodall and Wilson. "Understanding" behavior needs to be expressed in proximal (physiological) terms, ultimate (phylogenetic) terms or ontogenetic terms, not by the errors of anthropomorphism, anthropocentrism or teleology. The methods of behavioral science apply to both fundamental studies and applications.

CHAPTER 2. ANIMAL FORM AND FUNCTION: INVERTEBRATES ARE ANIMALS TOO.

Behavior is a product of an animal's anatomy and physiology. Without the right set of muscles and skeletal levers, an earthworm cannot run or fly anymore than an amoeba can. Nevertheless, earthworms and amoebae have substantial repertoires of behavior. There are certain needs of all animals which must be met by their behavior, and they find ways of meeting these needs as allowed by the anatomy and physiology basic to their kind.

The behavioral needs to which all animals must adapt can be grouped as: obtaining materials and energy, avoiding becoming materials and energy for some other animal, maintaining the right habitat, removing disturbing stimulii, avoiding the dangers of inactivity, interacting with conspecifics, and perpetuating oneself. Animals converge on adaptations to these basic needs for behavior since in some cases there are surprisingly few ways of adaptation, e.g. orderly interaction among animals which live in groups is often mediated by social ranking, or territory holding, or both. Adaptations such as these have been achieved time and time again. Mammals, birds, fish, cephalopods and insects all have produced territorial or rank-ordered species. The array of animal forms is so diverse that we must hypothesize independent origins of behaviors achieving these basic needs by many different genetic arrangements determining structures and chemistry, e.g. the behaviors are polygenic. Only in this way can we allow for the coming and going of complex, similarly functioning, behaviors down a line of descent. The alternative is simple Mendelian genetic determination of behaviors, for which there is little evidence (see Chapter 5).

Accordingly, we need to know how behaviors correlate with anatomy and physiology in the sense of what behaviors can be achieved by animals with different basic structures and chemistry. Fortunately, we have well established schemas for classifying animals according to their anatomy and hypothesized phylogeny. We can draw on traditional well-known classifications by phyla (Table 2.1) since the phyla represent basic differences in body plan. We can arrange these phyla in some way relevant to behavior. To do so we

Table 2.1. Animals arranged by major groups related to their behavioral capability (see Figure 2.1).

Major group and subdivisions	Common names	Common forms
Deuterostome section		
Phylum Chordata	Chordates	
Cephalochordata	Lancelets	Amphioxus
Urochordata	Tunicates	Sea squirts
Vertebrata	Vertebrates	Fish, amphibia, reptiles, birds, mammals
Phylum Echinodermata	Echinoderms	Sea stars, brittle stars, sea cucumbers, sea urchins
Other phyla, e.g. Chaetognatha	Arrow worms	
Protostome section		
Phylum Arthropoda	Arthropods	
Crustacea	Crustaceans	Lobsters, shrimps, crabs
Insecta	Insects	Butterflies, bees, caterpillars
Arachnida	Arachnids	Spiders, scorpions
Myriapoda	Myriapods	Centipedes, millipedes
Phylum Annelida	Annelids	
Polychaeta	Polychaetes	Marine worms
Oligochaeta	Oligochaetes	Earthworms
Hirudinea	Hirudines	Leeches

Table 2.1. continued.

Major group and subdivisions	Common names	Common forms
Protostome section (cont.)		
Phylum Mollusca	Molluscs	
Bivalvia	Bivalves	Clams, oysters, mussels
Gastropoda	Gastropods	Snails, slugs
Cephalopoda	Cephalopods	Squids, octopuses
Other phyla, e.g.		
Rotifera	Rotifers	
Ectoprocta	Ectoprocts	
Pogonophora	Pogonophores	
Lower Invertebrates		
Phylum Nematoda	Round worms	
Other "minor" phyla, e.g. Rotifera	Wheel animals	
Phylum Platyhelminthes		
Turbellaria	Flatworms	Planaria
Trematoda	Trematodes	Flukes
Cestoda	Cestodes	Tapeworms

Table 2.1 continued.

Major group and subdivisions	Common names	Common forms
Lower Invertebrates (cont.)		
Phylum Cnidaria		
Hydrozoa	Hydrozoans	Hydroids, siphonophores
Scyphozoa	Scyphozoans	Jellyfish
Anthozoa	Anthozoans	Sea anemones, corals
Phylum Ctenophora	Ctenophores	Sea combs, sea jellies
Phylum Porifera	Sponges	Sponges
Protista (Eukaryotes)		
Protozoa	Protozoa	Amoeba, Paramecium
Protophyta	Protophyta	Volvox, Euglena
Non-Eukaryotes		
Bacteria	Bacteria	Sewage bacteria, oil bacteria, sulphur bacteria

must modify the traditional phylogenetic tree indicating evolutionary relationships. By showing at what level of anatomical and physiological complexity particular adaptations to basic ecosystem demands appear, we gather information through seeing patterns of behavior and structure which can lead to understanding proximal and ultimate causation of the behaviors.

A convenient model of structural-behavioral relationships in animals is that of a multi-story parking lot (Figure

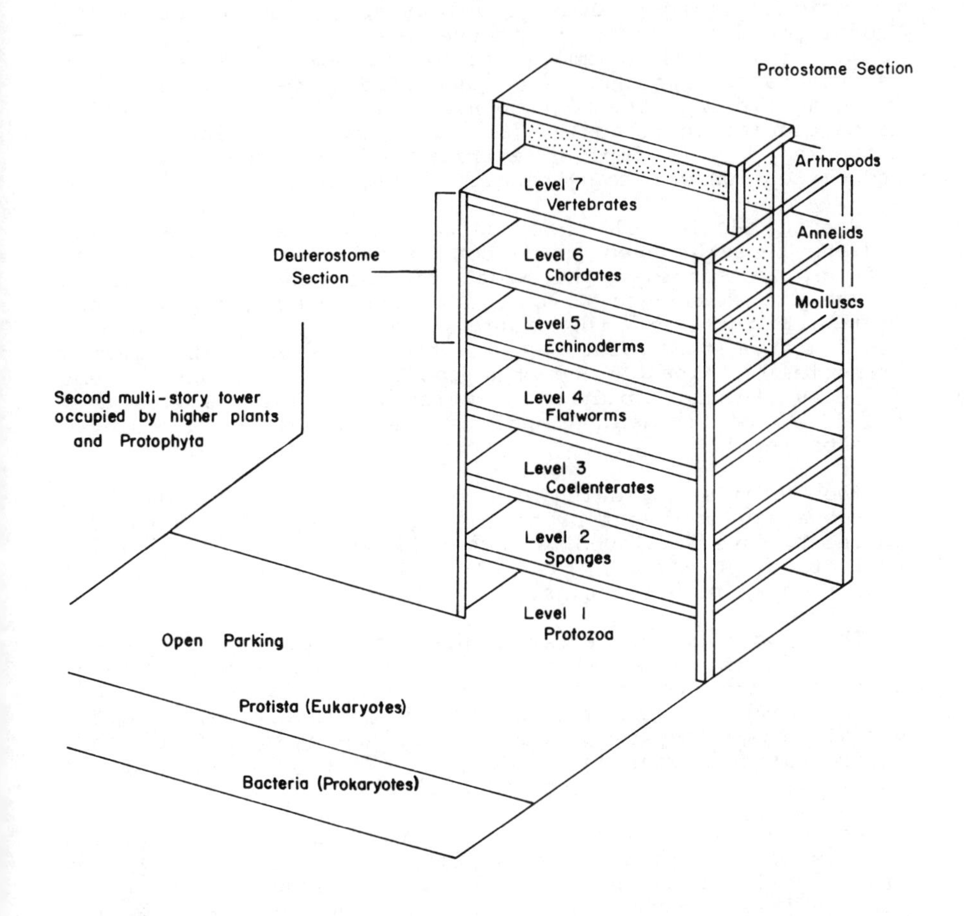

Figure 2.1. Multi-story parking lot model grouping animals at levels of structure which determine significant behavioral capabilities.

2.1). The phyla of animals are parked at progressively higher levels according to the presence of structures which allow more complex behaviors. Thus the lowest level is populated by Protista, with connections to outside parking shared with Bacteria, and to an adjacent smaller multi-story parking lot for plants (since they have some behavior). The three uppermost levels are divided into two non-connecting blocks reflecting parallel developments in Deuterostome and Protostome lines. The penthouse restricted parking at the top is open to the mammals and birds, some fish, and a few invertebrates which have acquired the sensory and coordinative capacity to recognize some of their own species as particular individuals, i.e, as a mate, a dominant, or an offspring. This implies extraordinarily fine sensitivity and perception with many behavioral implications.

Within each level of organization the structures which mediate behavior need to be considered, and to check these off we can consider the model of the behaving animal in Figure 1.5. We need to consider structures and physiology from all three of the model's compartments -- sensory, coordinative and motor -- and the connecting information transmission lines. But the capability of these systems reflects the basic body plan, which in turn is derived from the level of tissue and organ system organization achieved at each level.

The details following can be verified in any text on vertebrate and invertebrate biology, except for Table 2.1 where I have followed a generalized taxonomic division reflecting behavior, rather than current ideas on evolutionary relationships.

The most important structures for the more complex animals in the upper stories are the muscular and skeletal systems since they determine locomotory ability, the brain since it determines the scope of coordination, and the visual sensory system since vision has been adapted into configurational sensing in ways not often achieved by other modalities such as hearing and smell.

At the lower levels where structurally less complex animals are parked, the important properties are the traditional phylogenetic concerns of segmentation walls, body cavities, two or three tissue layers, multicellularity, colonialism, and intracellular complexity. Many of these lower level animals are highly successful forms present in large numbers worldwide, in spite of their more advanced predators and competitors.

The time scales for animals to reach the various levels must not be taken as constant. There is some argument that a body cavity and segmentation were achieved together, for example. The diagrammatic arrangement also ignores the route or routes from one distinctive behavioral level to another, i.e. whether each level was monophyletic and achieved by only one route or convergent and polyphetic, achieved by several routes, e.g. multiple body cells from flagellate and ciliate stems. The conceptual model also ignores the possibility that one level, the three layered acoelomates, could be a blind end accessed by descent from above. The routes and rates are not important in this context.

Upper level adaptations have frequently permitted novel converging on niches occupied or strategies adopted by earlier appearing forms, e.g. the current capturing strategy of tunicates and sponges (Vogel 1978). I am calling this reversional convergence to make the point that animals can evolve adaptations allowing them to recycle back into a highly competitive niche, in this case sessile rock dwellers. This is in contrast to the appearance of new major structures such as skeletons, which open up new strategies such as ambush or run-down hunting. In such cases (vertebrates and cephalopods) I am calling the convergences opportunistic. Some habitats such as sand interstitial spaces will have experienced an initial wave of opportunistic convergence by microorganisms, followed by reversional waves as more advanced metazoa evolved suitable microscopic adaptations, e.g. limited cell numbers in the rotifers.

The concept of strategy is used by biologists in ways peculiar to themselves. The terms carnivorous and ambush strategy show this. Ambushing could be called a tactic of the carnivorous strategy, but ambushing in turn can be achieved by the "motionless" and "hide" strategies (or tactics). Biologists, however, rarely use the term "tactic". They use "strategy" instead at many levels within an almost endless hierarchical nesting of adaptations to meet more generalized objectives. "Strategy" and "tactics" are the two parts of the concept that a behavior can be considered as having an adaptive objective, and there may be various ways by which that objective can be reached.

An example and summary for this introduction is given by the information transmission systems which have appeared in animals as an adaptation to the need to transmit ecosystem information from sensors to motors (see Figure 1.5). A number of such systems have appeared among animals and these are modelled in Figure 2.2. The differences are expressed in speed of transmission, which is the parameter of most behavioral consequence, but they represent substantial differences in structure and physiology. The systems range from the intracellular axoplasmic transport of Protozoa through at least two intercellular systems in the sponges, to the epithelial systems of hydrozoans and tunicates, on to the neural systems of most Metazoa. Plant systems, as in the bending leaves of *Mimosa*, are intermediate. The neural systems of Metazoa show a variety of structures and transmission speeds ranging from the low velocity nerve net of coelenterates, to the convergent adaptations for higher velocity, i.e. the giant cells of cephalopods and the myelinated nerves of vertebrates. There has been continuing

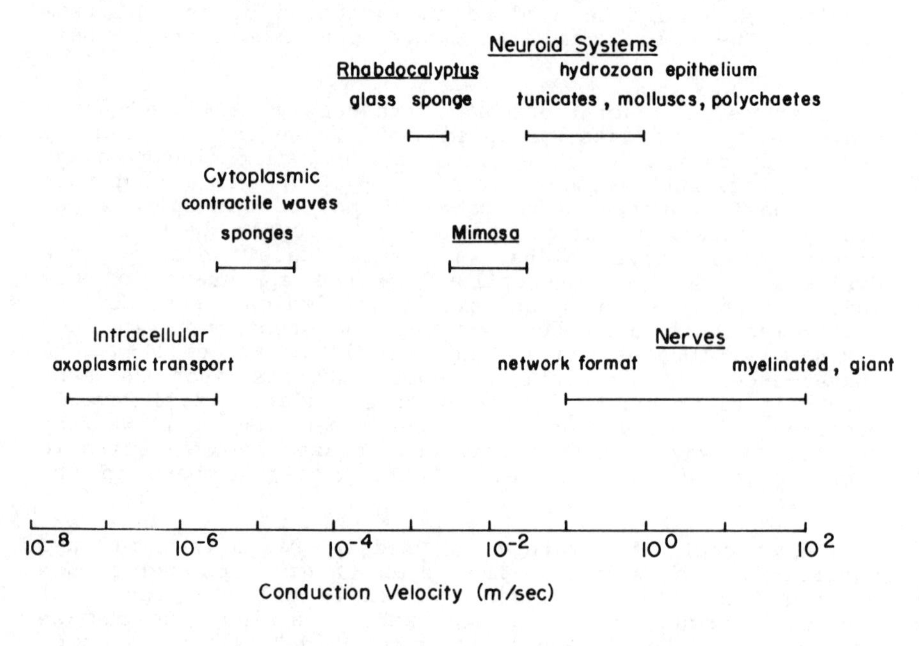

Figure 2.2. Speed of information transmission in various systems. (Modified from Mackie et al. 1983.)

selection pressure for animals to evolve information transmission systems and a number of such systems have appeared. Neural systems based on electrical processes have best met the need so far, to the extent that they produced relatively fast transmission, thus allowing the corresponding evolution of large animals and fast predators. In addition, complementary vascular systems have developed, allowing the transmission of information by flooding the body with chemicals, produced in vertebrates by at least some discrete endocrine organs, but in other animals by more or less distinct cell areas intimately associated with the neural system.

Vertebrates and arthropods: animated levers.

At the top level of the multi-story parking lot (Figure 2.1), including the penthouse, are arrayed those organisms which by acquisition of a skeleton have allowed the bunching of muscles to operate the animal body as a system of animated levers. This allows rapid movement (and large bodies), and opens up speed and distance-travelling adaptations only achievable occasionally in the lower levels of organization. The constraints on body flexibility arising from the presence of a skeleton and the inevitable repetitiveness of action by jointed levers also noticeably correlate with the development, at times bizarre, of stereotyped behaviors. These Fixed Action Patterns (FAPs) allowed behaviors to be ritualized into displays with information content such as threat, or subordination. These animals also have the sensory capability and brain power to perceive the sensory configurations and process the information into relevant decisions, some innate and some learned.

The adaptation has been achieved twice, in the vertebrates and the arthropods.

The vertebrates did it by means of an internal (calcareous) skeleton with a central core of an articulated backbone, on which are based two pairs of extended, jointed appendages. The latter have had extraordinary adaptability, ranging from the paddles of fish to the legs of fast running herbivores and carnivores, to the wings of bats and the hands of humans.

The arthropods adapted via a very different, even if convergent, route. The skeleton is external, of different organically-derived stiffening material (chitin), but jointed in such a way as to allow some body flexibility and the attachment of paired jointed appendages. These last vary in number, from the 10-20 plus of crustacea and myriapods, to the eight and six walking legs of the arachnids and insects respectively. The arthropod appendages may be specialized for various functions so that only a few are primarily for locomotion. They include sensors such as the antennae, feeding-respiratory-current organs such as the mandibles and maxillae of many crustacea, and mating modifications.

Some external structures of the arthropods, such as the intromittent organs of the spiders, and sensory hairs, have a separate origin from the paired appendages. External structures with various functions have often appeared among skeletonized forms. They include the hair and scales of vertebrates as well as the structures of the arthropods.

The leverage systems of both bodies have been accompanied by other distinctive features.

An important associated feature is the arrangement of the nervous system. In both groups there is a central core passing information along the length of the body, with a coordinating center close to the sensing front end. But there the resemblance in form ends. In vertebrates the core lies within the protection of the backbone, hence is dorsal, and ramifies to sensors and motor organs by elongated nerves. In arthropods the core lies ventrally within the skeletal frame close to the locomotory and many sense organs, and is divided into sections by minor coordinating centers, the segmental ganglia. Thus in arthropods the potential for central coordination has missed the opportunities provided in vertebrates, and taken to its present extreme (in humans) of enormous learning potential, information retrieval and comparison, language, reasoning and self-awareness.

The combination of limited central coordination and an exoskeleton which allows growth only by the dangerous procedure of moulting correlates with limits on size among the arthropods. The largest arthropod on land is a beetle 13 cm long, and in water lobsters and crabs have torsos that are not much longer. In contrast vertebrates, e.g. dinosaurs and whales, have grown, and the latter still grow, to about 30 m length. The vertebrate systems of the internal skeleton, jointed appendages and nerves/brain have allowed substantially greater size to be achieved, whereas the arthropod system has led to other specializations.

Of these, a new appproach to ecosystem demands arose with the development of flying organs. Flying animals can move rapidly through the relatively ambush-free environment of the air, and the dessication-resistant exoskeleton of insects has leant itself to such modifications in producing varieties of wings. Spiders have achieved flight less directly by gossamer strings. The convergent birds and bats have adapted paired appendages as wings, while other mammals use the same appendages for the lesser gliding and canopy leaping.

One of the most surprising adaptations of the many among the arthropods is a return to the widely-used strategy among the lower phyla of colonization. This has been accomplished in a significantly different format based on the insect form and size. Two major classes of insects (termites and ant/bees/wasps) have become colonial in the sense that grouped individuals show a division of labor which is often accompanied by structural and genetic adaptations. This adaptation has been carried to such an extreme that the termite hill, ant nest, or beehive has been equated with the body of an animal (Marais 1971) in the same way that the body of a multi-zooid free-swimming colonial siphonophore behaves as a single organism. Among the colonial insects the component units are individual animals comparable to other insects. This parallels the coelenterate zooids which can either be adapted to a separate life, for example as sea anemones, or colonial life as corals. The problem with this organism concept of an insect colony is lack of a physical information transmission system coordinating the activities of the individuals. There is evidence that this is achieved by chemicals carried or released by individuals, or pheromones permeating the living space within the external casing, possibly to a level of signal modulation, i.e. odor waves (Wilson 1970).

Another adaptive strategy retained widely among the arthropods is that of sequencing growth stages so that early and late do not compete in the same trophic niche, and may even occupy completely different ecosystems. Thus early and late may utilize different foods (mosquitoes), lates may not even feed but be adapted purely as breeding stages (some butterflies), and earlies may be larvae adapted as water column plankton or grubs rather than seabed inhabitants or fliers respectively. In the last strategy, there have to be extra adaptations assisting survival during the ecosystem change stage when there may have to be a migration or inactivity period (e.g. pupation) or both as the body reforms itself into structures adaptive in the next phase.

Among birds and mammals, and to a lesser extent among the lower vertebrates but apparently rarely among the arthropods, there is an adaptation with particularly important significance for behavior and social living. These animals can be placed in the restricted parking penthouse of the structural model (Figure 2.1). This adaptation is the advanced learning capability of recognizing other conspecifics as particular individuals, not just as other members of the species. Responding to one's species is widespread and is an almost essential adaptation for sexually reproducing animals. Individual recognition is by no means essential, but merely more efficient in terms of energy and time expenditure. When animals which live in groups have the capability of recognizing other individuals, then each time they meet they do not have to go through the entire repertoire of behaviors which determine their relative roles within the society. In particular, they do not have to go through the procedures of ranking, whereby a dominant has priority over a line of ranked subordinates. If they can recognize each other as individuals and learn relative ranks, they need not waste time and energy going through the whole process of winning and losing priority on each occasion when orderly priority for limited resources of food, space or mates is adaptive. Thus some fish may distinguish mates, but only mates, from all others, whereas elephants have been estimated to be capable of recognizing up to 500 individuals (Douglas-Hamilton and Douglas-Hamilton 1975). Humans can probably do better, at least sequentially if not at any one time. The ability seems to be based not only on the brain capability to store the complex configurations of stimulii characterizing different individuals, but visual acuity to distinquish the subtle differences in shapes, postures and behavior. For some species, e.g. dogs, individual odors may be the distinguishing feature.

Arthropods have not often evolved individual recognition as a strategy for group living. In societies of many individuals the chances of repeat meetings may not be great enough to make individual recognition by, for example, a guard, a more useful adaptation than recognizing a caste, e.g. a feeder, or a non-member of the colony.

Body segmentation

The level below the top in the multi-story parking lot model of behavioral evolution is occupied by the two minor groups of animals that converged on the possibilities opened up by segmenting an elongated body. In the vertebrate line of evolution there was one group of non-vertebrate chordates, and in the arthropod line, the annelids.

The backboneless chordates are a small group of largely insignificant creatures, one member of which shows the benefits of a segmented body. These are the Cephalochordata (the lancelets), which use a segmental arrangement of muscles around an internal stiffener, the notochord, as adaptations to burrowing through their marine sand habitats (Webb 1976). The other chordate group, which has not achieved segmentation, shows a different phenomenon, that of converging on life strategies already occupied by less complex animals. These tunicates, or sea squirts, have converged on the behavioral strategy of bringing the environment to the organism, i.e. by developing circulatory systems channeling the ambient medium through cavities surrounded by the organism's own body (Vogel 1978). It is as though these animals have grown around a space, through which they proceed to direct their fluid habitat either by capturing currents, or generating them by a pumping system. They extract oxygen and food particles and possibly also nutrients from the circulating fluid, and excrete their body wastes to it. This pump/capture strategy (Figure 2.3) was adopted very early in animal evolution, and even such late appearing and advanced organisms as chordates have taken structural opportunities to adapt to it (reversional convergence). They have also reverted to physical colonialism in which body units perfectly capable of functioning independently have coordinated themselves into a multi-unit superbody (see Section 2.6).

The annelids best show the advantages of dividing one's body into segments. There are many species, which are divided into the three main groups of marine polychaetes, largely freshwater and terrestrial oligochaetes, and the freshwater leeches. They have attracted diverse scientists and investigations so that the body forms and functions are

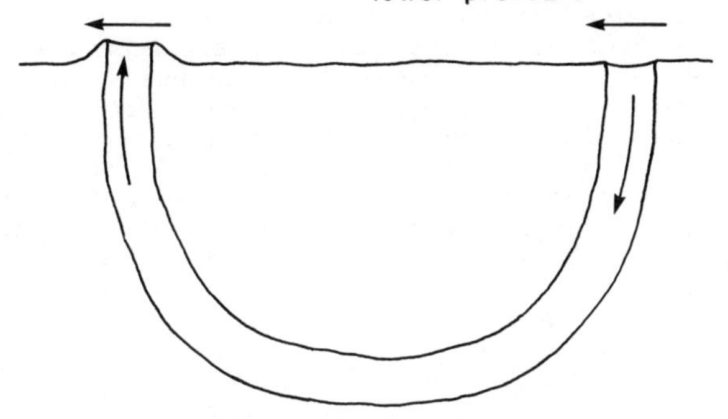

a. Burrows - sedentary polychaetes

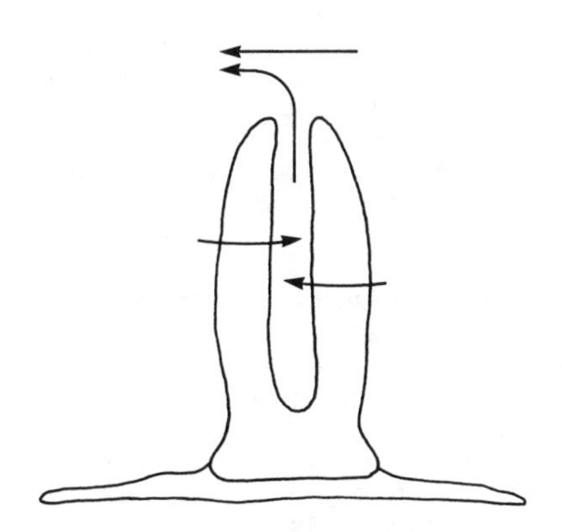

b. Whole Animal - sponge

Figure 2.3. The current-capture strategy for heterotrophism has been converged on by many different animals with either anatomical or behavioral adaptations and powered by (a) inherent physical forces with or without (b) backup support of a pumping system (Vogel 1978).

well-known (e.g. Clarke 1964).

The most effective adaptation for dividing the body into segments is called metameric segmentation. This term means that the body is divided into segments, each with a similar set of structures such as muscles, excretory organs, and reproductive organs. In practice, there must be differences since there have to be terminal segments at front and rear ends, the former needing to be loaded with sense organs. There must also be a feeding system, and the most efficient of these is a one way conveyor belt, so that again there must be terminal structures - a mouth for intake and an anus for ejection. Segmenting the body into compartments or repeating structures has occurred occasionally in other simpler invertebrates but their potential for modularizing the body has not been fully realized.

What serial repetition of segments allows is, for a moderately large animal, an elongated hydrostatic skeleton in which layers of concentric and longitudinal muscles can generate forces against contained body cavity fluids, and by appropriate levers provided in the form of setae or paddles can generate forces against the surrounding medium. With this adaptation to locomotion, the annelids have become effective paddlers and burrowers in the sea, freshwater, and on land in soil. In addition, some have converged on the current capture/pump strategy, so that many tube-dwelling polychaetes can live deep within an anoxic seabed pumping aerated interface water down where they can use it. The active paddlers and burrowers have taken the opportunity provided by speed (relative to other invertebrates) to adapt to the predatory way of life.

In addition, body segmentation was a key adaptation on which the later vertebrates and arthropods would build their own special adaptations. Both groups show the significance of segmentation, without which they could not have evolved. In the vertebrates, the serial arrangement of vertebrae and nerves of the central nervous system has some body flexibility and determined nerve supply patterns respectively. In arthropods, the serial arrangement of nerve ganglia is a fundamental constraint reflecting their limited potential for centralized coordination in a brain.

Body segmentation has been a key adaptation in the evolution of animals, allowing new convergences on basic niches and behavioral strategies to meet the demands of those niches. This applies to moderately sized animals since small animals with body cavities do not need segmentation for hydrostatic control. A few chordates and many annelids retain the simple forms of metameric segmentation and have not advanced to the greater diversity opened up by functional innovation in body structure.

Body cavities

The next level down in Figure 2.1 is the base where the two major lines of complex body forms first appeared. The common structural property between Deuterostomes and Protostomes is a body cavity. It is achieved, however, by different embryogenic mechanisms.

Below this level the invertebrates have largely solid bodies, or usually any body space is a current capturing pump-house system in which the body has grown around an outside space.

At this level, there is an internal body cavity formed within the middle of the three concentric tissue layers of the body (the mesoderm). This is most usually a true coelom, or coelomic cavity. It is called a true coelom since a few minor phyla such as the Rotifera and Ectoprocta have what are called pseudocoeloms that are formed by some other embryogenic mechanism, or located between tissue layers, or both. Above this level, the coeloms occur throughout Protostome and Deuterostome lines and their presence correlates with body plans. It is noticeable, for example, that a one way conveyor belt gut, which allows speedy processing of food, appears at this level, as do vascular (chemical) transmission systems. Both contribute to larger body size, speedier responses, and more power.

Where there has been convergence to a body form such as the presence of internal cavities, the form must have considerable adaptive value. The current capturing function of pump driven internalized ecosystem spaces is obvious. Much less obvious, however, is the survival value of a true internal space. Its presence, correlated with both advanced lines of descent, indicates that it has been a key base property for evolution of complex organisms. This is supported by the kinds of organisms at this level which show only the presence of body cavities without the advances of segmentation and its consequences.

On the Deuterostome line at this level, the main phylum is the Echinodermata, which encompasses the radially symmetrical sea stars, sea cucumbers and other forms. Generally these are relatively sedentary marine deposit or suspension feeders, often with protective shells and of medium size. They have specialized the body cavity into one or more fluid systems for internal transport or support, although a few have managed to achieve some speed and can function as scavengers or predators, e.g. brittle stars and sea stars. A few have converged reversionally on that ancient structural-behavioral strategy of bilateral symmetry with a distinct sensing forward end. They have done their best with what they have, but their distributional-behavioral limits are obvious even when they are not obviously dependent on their body structure. For example, the echinoderms are all marine. Why they could not adapt to the osmotic demands of freshwater as could the higher level, and many lower level phyla, is not obvious.

On the Protostome line, the main phylum is the Mollusca with its diversity of classes. Once again the base form as shown by bivalves and gastropods is sedentary, with protective shells. Gastropods have more mobility than bivalves and can be efficient herbivores, feeding on attached plants. A few are even predators taking even slower prey although the armed cone shells are lethal to unsuspecting resting fish. The cephalopods, however, have uniquely overcome their structural limitations and are the one group of animals without metameric segmentation which have achieved the strategy of fast movement, and thereby converged with vertebrates and arthropods as speedy and far-ranging predators. Cephalopods achieve their speed partly by a skeleton, which is usually internal although not jointed and provides a rigid support on which muscles have been inserted. But they also have a jet propulsion system based on current capturing (a novel function of the strategy) with paddles and a set of telescopic-flexible arms. In addition, they needed and evolved a fast transmitting information system, visual acuity, and a centralized coordinating system. They got them, but in ways completely different from those of the vertebrates and arthropods. The fast information transmission is achieved by individually elongated and wide nerves (giant nerve fibres, not myelinated as in vertebrates). The paired eyes look extraordinally like those of vertebrates. They have comparable acuity, three dimensional and some configurational vision, but are derived embryogenically in quite a different way. The centralized coordinating system gives considerable learning power (Wells 1978) and some social organization. Brain plus skeleton and limits to body flexibility allow some stereotyped display behaviors (and color changes) as in vertebrates and arthropods. With this package of adaptations, the cephalopods have produced some enormous species (the giant squids), some top carnivores within the marine ecosystem (octupuses), and some species with complex but largely unravelled territoriality and social organization (again the octupuses). A few slugs and other gastropods have gone part way with complex synchronized mating behavior, rarely appreciated since it needs time-lapse photography to exhibit its slow grace.

The body cavity (whether true coelom or not) in some of the minor phyla functions as a hydrostatic device allowing body movement (e.g. nemerteans, vestimentiferans) or extrusion and recovery of a proboscis or introvert (e.g. Echiuroidea, Priapulidae). Some compartmentalization of the body, in larger forms, may be present helping generate the necessary forces.

Three tissue layers

The next level of the parking lot for behaving animals is where animals show a head end and bilateral shape. They are adapted to moving forward and maintaining a constant direction, i.e. they can seek a stimulus source and not just aggregate (see Chapter 8).

This level is where the evolutionary novelty of three tissue layers is placed. "Tissue layers" refers to the arrangement of cells in developing embryos into two or three distinctive layers, which in turn are destined to become specific parts of the growing body. It is generally believed that the two and three layered embryonic forms represent early stages in the evolution of multicelled animals. Presumably there were many unknown, and unfossilized animals, representing simple and innovative ways of being free-living but slow moving herbivores and carnivores, since the tissue adaptation allows development of specific organs such as guts, gonads and ganglia in ways not achieved by non-tissued animals. There are many theories on how tissues evolved from non-tissued animals.

Three tissued or triploblastic animals consist nowadays of worms in the phylum Platyheminthes. The generalized arrangement is three concentric layers of tissues, the outermost and innermost being relatively thin and functioning as external (protective) and internal (digestive) surfaces. The intermediate layer, the mesoderm, is molded into a variety of organs serving information transmission and breeding functions but not chemical transmission systems, which appear not to be needed. The mesoderm of flatworms gives an indication of what is later achieved by the introduction of body cavities and segmentation. Three layers was a step which eventually produced complex animals, but the locomotory organs of muscles or cilia, or both, were not powerful enough at this level to generate rapid movements in large animals. They were, however, able to evolve into compound ciliary structures allowing novel feeding and swimming, which in turn permitted larval specializations and separate niches from adults.

The flatworms show two reversional convergences with behavioral consequences.

The absence of a circulatory system, even with the presence of the middle mesodermal layer, forces a thin body which can only be extended in two dimensions to allow the diffusion of nutrients, oxygen and wastes. One dimensional extension into a long narrow body is a form with advantages to a parasitic way of life. Two lines of descent, the flukes and tapeworms, converged on this niche. The niche must have been open as soon as multicellular organisms evolved, and is limited to smaller forms than the hosts. Thus initially it was probably occupied by the lowest level animals, the Protozoa, but the Platyhelminthes were able to compete for the niche and diversify in it.

Parasitism in turn required other behavioral-structural modifications to adapt to the problems of breeding in island habitats of living hosts, who are subject to the environmental impact of overpopulation, e.g. disease and death.

Many parasites have converged on the strategy of complex life cycles with larval stages occupying different ecosystems, and resistant stages (often free-living). These adaptations occur among parasitic and free-living lower level forms where the latter occupy environmentally hazardous habitats such as those having freezing or drying conditions. The behavioral problem for parasites in general, and larval stages, is to find and settle on the right habitat for the next life stage. The adaptive tactics are either to produce resistant forms in large numbers and distribute them so that the probability of accidentally getting into the next right set of conditions is high, or to adapt time-dependent sensory and locomotory mechanisms which give settling larvae some choice in their habitat seeking activities when they are more or less in the right place.

The Platyhelminthes deploy these adaptive tactics particularly successfully as parasites, and they do so with very little in the way of body organs other than what is needed temporarily at any one life stage.

One feature of flatworms has had an interesting scientific consequence. The lack of a bleeding circulatory system and a pain conscious brain makes some of the free-living forms particularly suitable for vivisection experiments. The major experimental thrust has been in learning capability related to simple bilaterally symmetrical and ganglionated nervous systems, and in memory regeneration and chemical transfer (McConnell 1974: 466-475).

Two tissue layers

On the lower levels of the multi-story parking lot are the cnidaria and ctenophores with a body base of only two tissue layers, the ectoderm and the endoderm. Two tissue layers do not limit these animals to two dimensional growth in length and width. They have achieved some three dimensional structure, and size usually expressed in radial symmetry. They have achieved this by what would seem an unusual adaptation, not the thickening of one or both layers, but by production of an internal non-cellular material, the mesoglea. On or through the mesoglea they have developed a nerve network system, occasionally with coordinative ganglia. Some species have even improved the network to include fast throughput channels which allow sudden contractions of body or tentacles as feeding or protective behaviors. Nevertheless they are rather limited in their size, muscular system, the leverage it can generate, and the speed with which it can be brought into action.

This structural limitation correlates with a number of adaptive strategies which allow the coelenterates to occupy the free-swimming, carnivorous, and suspension-feeding niches within the sea, all of which require some ability to move around, i.e. some ability to behave. They have been sufficiently successful that quite large forms such as the big jellyfishes, corals, and sea anemones exist, and some of these can protect themselves very effectively against much larger and more agile animals such as ourselves.

A protective and feeding adaptation has been the evolution of stinging organs, the nematocysts, which can be grouped into batteries, and are provided with toxins of varying power in different species. Armed with these weapons certain jellyfish, the sea wasps in particular (Halstead 1965), are effective contact predators, and can produce escape responses in larger animals, especially those of us with the learning capacity to recognize dangerous cohabitants.

A further adaptation is not quite unique since it appears by reversional convergence in some minor forms at higher levels such as the ectoprocts and tunicates. Here, however, it has been developed in unusually large mobile and sessile organisms. This adaptation is the strategy of colonialism, which coelenterates have achieved by actual fusion of individual zooids (each equivalent structurally to an individual sea anemone, hydra, or medusa). The fusion may be by interzooid extended connections as in colonial Hydrozoa such as Obelia, with or without specialized zooids adapted to specific needs such as reproduction. But fusion may be more complete, as in corals and Siphonophora, where zooids merge within an overall frame or flexible body respectively. The coral strategy of an overall frame encompassing the zooids places constraints on its potential, and they are sessile with little behavioral ability other than open and close, and pump and capture actions.

The free-swimming strategy achieved by such colonies as the Siphonophora and based on zooid fusion, has a fascinating connotation when compared with free-swimming multicellular and non-cellular animals. It seems that this strategy of functioning as a free-living animal with a substantial behavioral repertoire to achieve the necessary ends of survival, growth and breeding has been converged on in three ways. This is in the sense of organizing the body into structural units and combining into the behaving animal compartments of sensors, transmitters, coordinators and motors (Figure 1.5). The coelenterates and a few others have adapted by combining zooids with a base body plan capable of independent functioning as an individual animal. Most animals have generated individuals based on cellular structural organization, but the cell organizational structure is also capable of functioning as an individual, in the protistan form. In this last structural type the structural organization is based on intracellular organelles. The basic structural compartments of the behaving model animal can be achieved at that level also.

It is noticeable that the two body bases of cells or zooids have both converged on radial symmetry rather than the bilateral symmetry of most more complex animals. Radial symmetry correlates with slow locomotion and may be an adaptation to little ability to turn fast and present an armed, sensing front end towards ecosystem stimulii, i.e. the whole of the radial surface has some sensitivity to the environment.

This three-tier hierarchy of structural organization which has been deployed by evolution to the strategy of being free-living leads inevitably to speculation why there are only three tiers. There is some speculation that intracellular organelles may be symbiotic bacteria or other previously free-living, non-cellular organisms (Woese 1981). We can predict that in future aeons evolution will combine multi-zooid animals into yet more complex free-living organisms. In behavioral terms it does not seem to matter what living units evolution builds on. Organelles, cells or zooids can be combined in many different ways to produce living, organized animals, all showing the coordinated overt activities that we call behaviors.

Multiple body cells

Just above the ground floor for organized animal behavior is a level occupied by the sponges. These animals are of no more than moderate size, and all sessile. This implies a limited range of behavior since they do not move around their ecosystem. In actual fact all that they do is manifest the current capture-pump strategy (Figure 2.3), capturing their aquatic medium and circulating it through a plumbing system of tanks, pipes, and processing cells (choanocytes). It is the latter which extract particles and nutrients, excrete wastes, and drive the system by pumping through the action of one flagellum per cell. The rest of the body cells are scattered thinly over the surface of secreted strengthening material of various sorts or between spicules of the material. The functioning of the canal system is supported by some sort of information transmission called neuroid (Parker 1919; Mackie 1970), about which very little is known other than it is dissimilar in origin and structure from the neural system of other Metazoa (Lawn et al. 1981).

The inability of these animals to organize their multiple body cells into tissues which undergo coordinated rearrangements during embryogenesis reflects the significance of tissue forming as a base adaptation for the animal strategy of heterotrophism, the obtaining of food not by synthesis but by seizing the production of others. This energy-materials gathering strategy is facilitated by locomotion, the essence of behavior. The sponges do not move. They do not have the necessary muscular system, and their activities are limited to pumping, which is partly driven by specialized individual cells coordinated by a slow, diffuse and dispersed information receiving and transmission system. The motor cells are organized only to the level of intracellular organelles (e.g. flagellae), reflecting the prior adaptations of non-cellular animals at the base level below.

No body cells

The base of the multi-story parking lot modelling the levels of organization producing behavior is occupied by those animals whose living bodies are structured as a single unit -- the protozoan eukaryotes. These single units with surrounding walls, coordinating nuclei and other inclusions, have essentially the same cell structure that has been combined into the bodies of more complex and larger animals. They have adapted to a diverse range of feeding, survival and breeding strategies. They initiated many of the behavioral strategies later converged on by multicelled animals.

Such protozoa such as *Amoeba* engulf whatever small particles come within their reach but others (*Didinium*) are active carnivores devouring moving prey. There are also pumpers (flagellates and ciliates) bringing in their liquid medium and extracting organic particles from it. In short, the Protozoa demonstrate a wide range of behavior based on intracellular organelles. Locomotion is brought about by flagellae, cilia, and contractile strands flexing the body, or the internal tumbling of ameboid movement. Information gathering is conducted by sensory organelles, and transmission is unhindered by cell walls.

The major behavioral constraint on these organisms is the limited (microscopic) size, structure and power that has been attained by the non-cellular body. In some non-cellular plants the cells can be aggregated and coordinated into limited spinning locomotion (*Volvox*). It appears that coordinated heterotrophism needs powerful structures which can only be achieved within a single cell unit in small sizes incapable of driving a large body. For convergent larger structures allowing greater size the animal body must be organized into cells and tissues. The master strategy of animals has been to adapt the eukaryotic body plan into a building block from which metazoan bodies have been constructed. For this body to function and survive, i.e. obtain food, oxygen and water, behavior is a necessary consequence.

The eukaryotic body form is currently believed to be the product of symbiosis between prokaryotic bodies (several kinds of non-nucleated, non-organelled free-living cells). Current prokaryotes are even less capable of behavior than the simplest eukaryotes, being largely limited to spinning swimming, orienting by kinesis aggregations (Chapter 8). All are microscopic autotrophs absorbing various forms of nutrients for chemosynthesis; none are heterotrophs with adaptations for capturing live food.

The multi-story parking lot model has its ground floor extending to a second tower by an open area. This reflects the existence of the autotrophic or synthesizing organisms.

The interconnecting area models the existence of the array of bacteria and non-cellular forms, which although they are of varying origin and structure (Woese 1981), in behavioral terms reflect the constraints of size and lack of the selective pressures of heterotrophism inducing varied adaptations to food gathering. Bacteria show some motility and sensitivity for specialized media in which they live, but respond largely by body division and the production of resistant stages rather than by behavior.

The model shows an interconnecting apron leading off to an indicated second tower which divides photosynthesizing plants into levels of structure and complexity parallelling those of animals. In behavioral terms this tower is of little interest, since only a few plants behave in the animal sense. Behavior among plants has been an unusual adaptation, not essential for the photosynthetic way of life, but occasionally bringing survival advantages to particular species.

Thus at lower levels within the plant tower, many species have motile single cell stages with the necessary organelles. There are migratory forms in water columns and mud flats capable of varying their depth daily within the habitat to ensure their being at the optimum illumination for photosynthesis. There are motile gametes and spores actively seeking other gametes or an appropriate habitat. There are peculiar aggregating cells of slime-fungi temporarily living together as motile organisms. At higher levels, with bodies arranged into the equivalent of animal tissues, some plants such as sundews can be provoked into the relatively fast action of capturing and digesting insect prey initially held by sticky secretions. There are lesser behaviors such as the leaf-bending of *Mimosa*. But in higher plants behaviors appear patchily, and are of little phylogenetic consequence in determining lines of descent through evolution.

Behavior is a base strategy of animals. It is adaptive to and diversifies the heterotrophic way of life. It rarely appears outside the animal kingdom, and especially not in the form of complex social behavior.

Habitat implications for composition and behavior of the fauna

The habitat of an animal to some extent predetermines its behavior. A biologist with an interest in the interacting behavior of species in an area, e.g. someone engaged in environmental impact assessment or food network analysis, can predict that species occupying certain types of niches will be present, and what their interactions are, e.g. predator/prey, competitors, herbivores, parasites, symbionts, etc. In this context, there are three major habitat divisions: the column, interface and base. These show best in the marine environment, where each is well populated, but the same divisions occur on land (Table 2.2) and in fresh water. The principles of biological pattern in present-day ecosystems should be considered since there have been substantial changes over geological time, e.g. no free oxygen, no land plants.

The water column in the sea is a virtually featureless environment without much structure other than that provided by the organisms themselves. Thus plankton occur in clouds which are detectable acoustically by man if not by the most advanced of the resident predators - the baleen whales. Physical chemical variation at levels detectable by simple animals is limited to vertical gradients or fronts of temperature, salinity, pressure and similar properties. Resident organisms are planktonic and occur in clusters which serve the function of maintaining the animals near

Table 2.2. Major habitat divisions and faunal composition.

Major habitat division	Sea		Land	
Column	Water column (featureless)	Plankton Nekton	Air column (features)	No plankton Mostly transients, flyers, gliders, etc.
Interface	Marine forests and meadows (structured)	Diversity of animals	Forests, grasslands, etc. (structured)	Diversity of animals
Base	Sediments (featureless)	Interstitial macrobenthic burrowers	Soils (featureless)	Inter-stitial burrowers

conspecifics for breeding purposes in a featureless environment, but increase their risk of being eaten if the cluster is found by a predator. A few large mobile animals occur, primarily squids, fish and cetacean and phocid mammals. These generally have the capability of high speed and long-distance movements and some follow narrow front migration routes. Some cetaceans may have a unique adaptation of maintaining very long-distance contact through a deep sound window. The featureless nature of the water column and its correlates of drift plankton, with only specialized carnivorous nekton, allows some prediction of the behavior of the component animals and the food networks that they can form in any oceanic or coastal region.

The air column is a very different habitat with its low density medium, and there is no aerial plankton based on drifting plants. The air column is populated only by transients which fly, glide, leap or otherwise propel themselves through space by their own flapping or parachuting, or with an assist from lightweight silk lifelines. All return to the physically structured ground at least for breeding, often for feeding, and usually for most of their needs. The extreme variant are swifts, which return for breeding and nocturnal sleeping. Some may even sleep in the air (Lack 1956) by flying upward at night, thus avoiding the complicated structures which project into the air from the base land below. Thus all aerial transients require the sensory, coordinative and motor capacities allowing them to pilot themselves through the complex structures jutting out from their land base. Very few have achieved this, and the species lie mainly within the insects and birds, with a few mammals and arachnids. There were some reptiles, and there are a few aerially transient fish. But these are restricted to the almost unstructured air over sea surfaces.

It is the solid interface characterized by forests and grasslands where physical structure has developed to its greatest extent due to plant adaptation, and given rise to an enormous diversity of animals with varied and acute sensitivity, coordinative and motor ability which enable them to make their living between, on and in the plant structures. This is the main habitat in which terrestrial behaviorists work, and they concentrate their efforts largely on the complex social animals, primarily mammals, birds and insects. In the sea there is a lesser array of such forms, and they have arisen primarily from different groups: fish, crustacea, and cephalopods, although some birds and mammals have managed to enter and compete. The marine behaviorist is able to draw parallels with terrestrial behaviorists' concepts and data base, and can expect the component behaviors of the social adaptations which have developed in land forest and grasslands. It is noticeable that deep marine forests and meadows are animal based, i.e. composed of sessile tunicates, bryozoans, coelenterates, etc., not plants. These animals provide the structure of the habitat, and have relatively limited behavior themselves compared to the resident epifauna.

The underlying inhabitable material of the base (sediment and soil) is relatively restricted three-dimensionally and virtually featureless, having only simple variations in environmental properties vertically and horizontally. In both land and sea, diverse populations and simple microscopic forms occur. In the sea an extraordinary diversity of large worms and a few molluscs, crustaceans, etc. have taken up burrowing adaptations in the soft bottom, but generally are in contact with and depend on organics rained from above, and do not maintain social contact with each other. On land, a few complex social insects, and even fewer mammals, have gone underground, and some specialized forms have raised the underground habitat around themselves in the form of hard colonial nests.

These patterns of organisms cohabiting are there and can be used in ecosystem investigations. They can guide efforts to understand ecosystem processes and the behaviors by which the processes are brought about by different species in different localities.

A phylogenetic summary

Animal behavior is an extension of protozoan locomotory ability which serves several functions. Heterotrophs can capture prey, gametes and conjugants can seek or chance upon other conspecifics, autotrophs can seek environmental optima such as illumination levels, and free-moving or sedentary forms can escape predators or environmental stresses.

Among Metazoa there has been a trend to large size, supportive skeleton, fast movement, novel weapons, acute sensitivity and complex coordination. Resulting complex behaviors still serve the functions of heterotrophy, survival and gene recombination, but adaptations serving one function may imbalance others and require corresponding counter adaptations if the individual is to survive. Thus the fast speed of an open water predator such as a fish may need corresponding social organization in order that conspecifics do not separate in a visually formless void, never to find each other again for breeding.

For any adaptation there are both benefits and costs, and the costs must be met in some way for survival.

The various phyla of animals can be arranged in an order reflecting behavioral complexity and progressive opportunistic convergence on new ecological niches and adaptive behaviors arising from the evolution of new body forms. Thus not until plants had attached by holdfasts and roots could grazing and browsing herbivores arise. Herbivory has been converged upon by such forms as molluscs and sea urchins in algal forests, and mammals and insects on grassland. Each development changed the nature of the ecosystem and opened new niches.

Progressively more complex structural stages with behavioral consequences can be ordered as non-cellular bodies, multicelled but non-tissued bodies, tissued bodies (two and three layers), cavitied bodies, segmented bodies, and animated levers (Table 2.3). The first three groups (levels 1-3) have a basically similar limited range of strategies open to them: slow free-moving herbivory, carnivory and breeding, with reversional convergence on sessile or sedentary current-capturing and multi-bodied colonialism appearing every now and then. Body cavities, segmentation and skeletal levers opened up new opportunities for increased size and speed. This led in turn to behavioral strategies of long distance habitat changes and ambush-chase carnivory. These locomotory adaptations to survival and heterotrophy were accompanied by another set of adaptations favoring both heterotrophy and sexual reproduction, i.e. social organization, Fixed Action Patterns and learning. The important base adaptation for

Table 2.3. Animal levels of structure determining significant levels of behavior.

Level 7	Brain	--->	Individual recognition
	Muscular-skeletal system	--->	Rapid movement
Level 6	Metameric segmentation	--->	Hydrostatic skeleton
Level 5	Body cavities	--->	Complex organs
Level 4	Three tissue layers	--->	Complex bodies
Level 3	Two tissue layers	--->	Cell organization
Level 2	Body cells	--->	Cell interaction
Level 1	No body cells	--->	Small size and organelles

socialization is quick recognition of conspecifics. This has been extended in higher vertebrates to recognition of some conspecifics as particular individuals -- mates, dominants or offspring.

Thus the essential behavioral consequences of animal evolution have been progressive increases in size, power, speed and distance of movement, a weapons arms race between predator and prey, and social organization, peaking with the ability to respond uniquely to particular individuals and other learned adjustments to ecosystem variability.

Summary

The behavioral capability of an animal is determined by both its form and environmental selective pressures. Its particular form lies in a sequence of evolutionary order from non-cellular bodies, through multicelled but non-tissued bodies, tissued bodies (two and three layers), cavitied bodies, segmented bodies and animal levers. Each stage in the sequence imposes limits on what can be achieved ranging from individualistic opportunistic heterotrophy of microscopic protozoa to the complex social lives of mammals capable of recognizing each other as unique individuals within an organized group. Each habitat has corresponding ranges of herbivores, carnivores, scavengers and parasites, etc., with convergent behavioral adaptations as well as activities uniquely adapting them to specialized niches. These patterns of form, function and niche can direct behavioral investigations in such applications as environmental impact assessment, animal domestication and conservation, and human phylogeny.

CHAPTER 3. THE ETHOGRAM: INVENTORY, CATALOGUE AND CHECKLIST.

When we observe an active animal, such as a duck on the local pond or a lion (on film), it appears to have an endless variety of behavior. Provided we take the time to watch, we often notice something new among the frequently repeated behaviors which make up much of its daily life. What is endless is actually the way in which the animal deploys its capabilities in new situations, or in situations that are slightly different from normal. Thus mallards which normally dabble for their food in shallow water, have a variety of dive and lunge tactics for taking food that is just out of their normal reach.

This apparent infinity of behaviors presents some problems in their study since it seems to preclude a traditional scientific method: cataloguing. Cataloguing is a foundation for science since it provides a list, ordered in some way, documenting phenomena needing understanding. The nature of the ordering can provide guidance to understanding causes. Cataloguing has been applied to science in many areas ranging from physics through chemistry to biology, and within these areas to subdivisions of each. In biology, taxonomic cataloguing guides scientists through the enormous diversity of species. Anatomical cataloguing provides a phylogenist with lists of structures for tracking homologies. Physiological cataloguing divides the animal body into organ systems for chemical and electrical investigation.

The apparent endlessness, or "infinite" variety, of behaviors does not preclude the cataloguing approach to their study. In fact, it makes it more essential. We must have a catalogue of the behavior of a species if we are to get a sense of its capability, and its adaptiveness to the ecosystem in which it lives. The problem is to arrange the catalogue into a hierarchy of inclusive high level categories so that all possible behaviors are nested in somewhere.

The practical value of such a catalogue is that it functions as a checklist. We can use it for guidance in reviewing the behavior of a species of interest, and in making actual observations of an individual animal, either in the field or laboratory. A comprehensive catalogue can guide us and direct our attention to capabilities likely to exist but not yet seen. This is particularly important, but easily overlooked, in site-directed environmental impact assessments (such as whether caribou will cross arctic pipelines), veterinarians' clinics (where pathological manifestations may displace the functional), and Skinner box training experiments (where an animal expresses the functional within the constraints of the box).

We call a behavioral catalogue an ethogram. We produce it by making an inventory of behaviors. We use it as a checklist. The three part approach of inventory, catalogue and checklist is fundamental in the behavioral investigation of a species. Eliminating it renders the investigator liable to miss significant behaviors not occurring during the period of observations. In this way, behaviorists come up against the ecologist's typical sampling problem. An ethogram provides guidance in stratifying one's behavior sampling so as to get adequate coverage of all activities. It helps to avoid biasing one's observations on the obvious and frequent parameters of the animal's world of activity, or the investigator's narrow terms of reference.

Since explicit formulation of the ethogram concept by early ethologists such as Tinbergen (1951), experience has shown that upper level encompassing categories can be conveniently established in functional terms. Behaviors can then be grouped into these functional categories, or contained subdivisions, so that a hierarchical nested system is developed. This lends itself to listing (Appendix 1), and to exposition in writing (Appendix 2), verbally, or by media presentations (Appendix 3) of film, videotapes and color slides.

A set of functional categories which can be used in this way is: obtaining materials and energy (largely feeding), avoiding becoming materials and energy for some other animal (largely escape), maintaining the right habitat, inactivity, removing disturbing stimulii, social organization, and perpetuating oneself.

These seven categories need a little stretching in the case of complex mammals with their versatility in dealing with new and complicated situations. As an example, play is a distinctive behavior easily recognizable in some animals, and arguably occurring in others (Fagan 1981). Play is often seen among young mammals and we can usually agree that what we see is play. It serves the function of practice and gaining experience. However, some adult animals also play. Among adults, play may pass the time (e.g. cats), facilitate social bonding (e.g. ravens), bring exhilaration (e.g. humans), or aid courtship (e.g. dogs). Adult play is difficult to describe so that others will agree that what has been seen is actually play. It is also difficult to determine the function of adult play at any one time. It might be considered a part of the ethogram categories of inactivity, social organization or perpetuating oneself. Many mammals and some birds appear to have converged on play as a tactic serving several strategies.

The difficulty of slotting a behavior such as play into a catalogue is a typical classification problem for biologists. It does not invalidate the method, but does complicate it. Natural phenomena can be so complex that some examples cannot be readily slotted into our categories because their limits were defined to catch the examples which are most obvious to us as observers from another species. The Onycophoran *Peripatus* defies classification as an annelid or arthropod. Community categories such as "ece" and "faciation" (Clements and Shelford 1939) need such care in their application to intergrading ecosystems that they have virtually dropped from use. Metal-binding low molecular weight proteins of invertebrates were not originally accepted as the metallothioneins previously found in vertebrates (Roesijadi 1981). Our arbitrary categories for classification purposes will always find difficulties with some biological phenomena.

The rest of this chapter discusses behaviors falling within six of the listed categories, and illustrates the diversity of actions serving the various functions. This chapter (with Chapter 4 following) can serve as a general guide to preparing an ethogram of the type shown in Appendix 2 (Pacific salmon, genus *Oncorhynchus*). A checklist is provided as a more concise guide (Appendix 1). One category, social organization, is reserved for separate presentation (in Chapter 4 following) since it spans a variety of behaviors between individuals, by which they organize themselves into groups achieving many different functions.

Obtaining materials and energy

Obtaining materials and energy is more than just feeding. We can see this by considering the animal body as a chemical factory which must import all its materials, process them into forms suitable for incorporation into its own body for growth, and extract the driving force of energy. It must then export or sequester unused or unusable wastes in ways which have minimum impact on itself or its inclusive fitness. All this activity must be done within a liquid medium, water, which is the solvent allowing the enormous number of necessary biochemical reactions to proceed.

In short an animal must import food, water, oxygen and any other compounds which the animal cannot synthesize itself, and export potentially dangerous wastes such as carbon dioxide, nitrogen, trace metals and organics. It imports its raw materials by eating, drinking and breathing, or by some combination intake behaviors. It exports most wastes by defecating, urinating or some combination of the two (as in birds). It removes carbon dioxide efficiently by linking the process with the intake system for oxygen. It sequesters compounds in inert biochemical form in tissues and skeletons largely by physiological processes rather than by behavior. Some species have secretory processes which export materials not falling within the usual categories. Thus highly saline fluids can be excreted as tears, and assist in osmotic control, keeping the body fluids of marine animals (e.g. turtles and sea birds) at concentrations which permit the biochemical processes of life despite the inrush of salts from the seawater medium.

Both import and export may involve behavior, such as feeding and defecating, and in addition the behavior or a product (shell, tears, etc.) may be adapted to a secondary behavioral function (protection, social signalling, etc.).

Feeding is accomplished by several strategies, of which herbivory and carnivory are the most obvious. Herbivores have a variety of strategies for feeding on plants and as a consequence cohabiting species need not be competing for food, i.e. they partition the food resource in some way. Common terms for types of herbivory are grazing and browsing. These have been used in such a variety of ways by biologists interested in different ecosystems, that generalized definitions are needed. In everyday use browsing refers to feeding on succulent canopy leaves and stems, and grazing refers to eating grass (which regrows if the rootstock is left) or entire low growing plants (which often do not regrow). There is obviously a significant difference in ecological impact between the two. Accordingly, browsing can be conveniently used for herbivory in which relatively small amounts of the food are eaten from each plant, which can then regrow. Grazing can be conveniently used for feeding which destroys the plant, or which seriously impedes regrowth. The latter can then be applied to ungulate grazing on dry African grasslands, sea urchin grazing on kelp beds, which can be destroyed by holdfast release (North and Schaeffer 1964), and even suspension feeding on phytoplankton by aquatic zooplankton and benthos.

Carnivores also have two base strategies, and these are refined by specializations arising from fast locomotion. Fast predators use the strategies of ambush or rundown (coursing), depending essentially on whether they are faster than their prey or not. There is actually a balance between the speed achievable and the distance it can be maintained. Many mammalian carnivores can combine the two strategies, depending on the prey available to them at any one time. The ambush strategy can be divided into hide or freeze (drift) ambushes. A lion stalking or waiting for an approaching zebra can creep stealthily under cover of long grass until the two are close enough for the final dash by the lion. A motionless, camouflaged snake can wait for small prey which would see it if it moved. The drift ambush strategy is also available to slowly moving animals. A quiet jellyfish can drift close to a prey fish until contact and quick capture. Sedentary or sessile forms must use other strategies. These are particle feeding and current capturing, both of which pass the aquatic medium through the body of the animal, and food particles are extracted and digested.

Particle feeding comes in two forms. Suspension feeders suck fluid from the drifting water column around them, and extract food particles on gills or other filtering or sticky organs. Deposit feeders vacuuum up a slurry of mixed organic and inorganic particles from the sea or lake bed around them. Both need some sort of pump to create the necessary suction, and muscles to deploy their intake pipe in the water or around the seabed. They may be aided by the more primitive strategy of current capturing, where the intake current arises from the physical effect of the relative positions of intake and output apertures (Vogel 1981 and Figure 2.3).

Internal parasites import food by simple diffusive processes not requiring behavior, but there is no strict dividing line between an external parasite and a small epifaunal predator, the animal-eating equivalent of a browsing herbivore. Some nudibranch molluscs, for example, can live on the bodies of larger sea anemones and other sessile forms, and feed on the outside tissues of their host-prey. Between such small epifaunal predators and attached parasites such as the embedded highly modified barnacle *Sacculina* there is a complete spectrum of symbionts, commensals and mutualists, feeding by one or another behavioral method, with more or less impact on and counter adaptations by their host-prey, or co-evolution by mutually benefitting species (Chapter 4).

Many animals are scavengers which feed on dead and decaying food. This requires behavior to locate (e.g. vultures soaring) and to compete with others (e.g. hyena pack aggression).

Other species may be omnivores quite capable of adapting feeding strategies to whatever is available. Chimpanzees, which are normally browsinq herbivores, will hunt small monkeys if the opportunity to catch one easily arises. They have developed social behaviors around the eating of such prey. The owner (capturer) dispenses (shares) in ways that are not seen with plant food.

Finally some animals, such as bees, when faced with choosing from a diversity of food sources, will pick and maintain their choice for some time. This choice appears to maximize the energy-materials benefit-cost ratio, i.e. energy obtained relative to energy expended to obtain, or energy required to seek alternatives.

As materials are either taken in as a slurry in a liquid medium, or must be digested and dissolved prior to processing, water is often required separately from feeding. It is imported by drinking. Drinking for many land animals is a dangerous business, since the process of putting head to water lessens the animal's chances of perceiving predators. So counter strategies are needed to balance the risks of drinking, e.g. reducing the frequency of drinking, or social sentry duties by guards. In some far-travelling oceanic birds, the need for freshwater drinking has been eliminated by an internal desalination system so that the albatross, for example, only needs to come to shore to breed, not to drink.

Oxygen must be imported for a variety of oxidative biochemical processes, the best known of which is the combination with carbohydrates to release energy as the driving force for the living animal. In aquatic animals oxygen may be obtained through the skin without behavior, or by special organs, gills, which may require appropriate behavior to flush, regularly or periodically. In terrestrial animals special organ systems are also needed, and these may require regular or periodic breathing (flushing of lungs or trachea).

The import processes may bring in unwanted, unusable or excessive materials. These are generally removed by the process of defecation which, like drinking, may be dangerous when in progress, and thus is rapid. Defecated wastes have the potential to carry pathogens capable of impacting the defecator or its group mates by infection. Some animals, such as group living rodents and cats, have developed fecal placing (toilet) behavior which minimizes the health risks, whereas others such as the gorilla can afford to be messy due to their transient habits. The hippopotamus with its feces scattering behavior (with its tail whirring like a propeller) has adapted the export process into a behavioral and chemical signalling system, advertizing its presence to other hippos. Bivalve molluscs and other particle feeders have a first line of rejection for unwanted particles, and produce pseudofeces by selection processes prior to ingestion of organics. Polychaetes may build such rejected material into species-specific tubes.

Removal of wastes such as carbon dioxide and nitrogenous compounds produced from the biochemical processes requires various behaviors. Carbon dioxide removal is commonly achieved by breathing, and nitrogenous waste removal may be more or less linked to defecation by urination. Some animals such as dogs keep the two separate, and have adapted urine as a social marker indicating their presence to other dogs. Breathing may be adapted as a social signal as in the bubble-blowing nest building of air breathing Siamese fighting fish. Breathing may be reduced in low activity states such as hibernation, and increased after or during high activity.

In summary, obtaining materials and energy by heterotrophy has imposed a need on animals to capture food. There has been a trend to progressively more complex capturing strategies. This may separate import and export of materials into several processes (eating, drinking and breathing; and urinating and defecating respectively), or they may be combined in various ways. Within each major materials gathering strategy such as carnivory, herbivory, and particle feeding, there are several tactics. Whichever is adopted has an impact upon other survival needs, so that counterbalancing tactics are needed and have been evolved. In many cases, materials gathering or waste removing strategies have been adapted into social signals (see Chapter 4).

Avoiding becoming materials and energy for some other animal

If animals must eat, some other animal or plant must be eaten. Nature is "red in tooth and claw", and many animals are eaten alive. Both plants and animals have developed several strategies to minimize this risk, if not the pain. Plants do so by passive adaptations, some of which are shared with animals, such as acquiring aversive tastes or odors. This is a long-term strategy, requiring predators to co-evolve parallel non-eating adaptations. When animal coordinating centers had evolved to the level at which individual learners could remember their aversive responses on their first attempts to eat that prey, this distastefulness strategy must have become more practical in the sense that even a slight trend towards an unpleasant taste to an intelligent predator would bring an immediate increase in fitness of the prey. The evolution of parental behavior and teaching increased the utility of the aversive taste adaptation even farther.

The base behavioral adaptations for survival, however, are active, drawing on fast movement. They consist of alert, escape, hide and/or defend oneself. "Alert" consists of stopping other activities and placing oneself to maximally receive stimulii indicating predator approaching. The escape response is an immediate fast muscular action which in locomotory forms removes the attacked animal from the danger point. It may consist of a violent dash, leap, swim or upwards flight allowing the animal to move violently, but in a relatively undirected way, away from a fast approaching predator, or away from a stimulus that has a high probability of being produced by a predator. The escape may be aborted to a startle response if no backup stimulii confirm that a predator is attacking. It may be accompanied by an alert call if the animal is a member of a group (which may also benefit from being alerted). These alert calls may have the peculiar tonal property of giving little clue to their direction. The animal may have evolved startle color patterns which flash and alert group members. These are often located at the body's rear end, which is what other group members would see as they explode apart or follow each other towards a common end point.

The escape response may also consist of withdrawal movements by which the most vulnerable, soft part of the animal is withdrawn into a protective device such as a shell. Thus clams and snails can withdraw relatively quickly into their shells, limpets and sea urchins can clamp down to their habitat and resist attempts to pry them free, and even soft animals without shells such as jellyfish can withdraw tentacles and form themselves into a more compact shape.

Accompanying such movement responses, there may be defensive behaviors which produce aversive responses in the attacker. Many animals can direct a jet of chemicals, usually from the rear end, but not always. Fulmars spit, whereas skunks and ants jet their chemicals from anal or other glands. As with aversive tastes, these adaptations become more effective from learning and parental teaching on the part of predators. Armed animals may defend themselves by horn, tooth, claw and leg.

The end point for escape behaviors can be either the aggression of self-defense, the follow-up of "hide", or the extended run of "flight". The dart must terminate quickly, otherwise the escapee goes into the slow movements of fatigue and increases its risk of being taken by the following, or another, predator. "Hide" consists of placing oneself where one is undetectable. This may be behind a screen of habitat materials, such as a physical hiding place, a produced screen such as the ink jet of cephalopods or the animal's own shell on which a trapdoor may slam shut (e.g. the operculum of some gastropods). But the hiding behavior may be more subtle. Many young birds and mammals "freeze" with or without displaying cryptic coloration which assists camouflage. Serving the same function as the "freeze", is a group response in which a flock of small birds or shoal of fish forms a concentrated temporary pseudocolony in which each animal unit maintains a minimum distance apart and the coordinated flock moves as a single unit. The selective advantage appears to be that a hunting predator cannot easily take a visual fix on a particular individual in the moving group and so cannot complete its strike. (Some fish predators have counter adapted and they hurl themselves into a dense shoal of fish, knocking some of the shoal out. These can then be caught easily.) The coordinating system for the flock to move as a dense group is generally believed to be virtually instant sensitivity and response to visual cues of movement made by the members of the group that are ahead.

The evolution of learning capability has had benefits to animals in their escape and hide behavior. Many animals explore new habitat, or regularly patrol it as though they are curious about contained locations and their potential to serve in various capacities. Lorenz (1952) described a water vole hesitantly patrolling its new cage until it had learned specific routes, and thereafter speedily traversing such routes, especially when alerted. Salmon can be seen passing through fish ladders by initial patrols of each step before leaping up to the next. The animal which has learned where its nearest hiding place is, can direct its escape dart there and not just rely on an explosive undirected flight away.

Finally, an animal must come out of hiding. This is often done slowly with cautious sensing of the ecosystem around it, and return darts back to hiding or withdrawal.

In the event that the predator wounded the prey, the latter may extend its hiding, thus reducing the risk of further attack while injured. It may engage in the antiseptic behavior of wound cleaning either singly, or with colleagues.

These behaviors serve the function of removing a prey species from a predator, and placing it in a position to reduce the risk of being caught.

Maintaining the right habitat

Animals live within an ecosystem, part of which is their habitat. This includes structures (which may be biological in origin such as the forest canopy) and environmental limiting factors such as temperature, which can freeze or overheat them. Habitat has a significance to animal behavior that is different from the significance of the ecosystem biological processes such as the interactions between predator and prey, and competing and mutualistic species.

Animals use behavior to maintain their right habitat. They do this in two ways: balancing limits and optima, and restricting movements.

Mobile animals may restrict themselves to only a limited space within the total that is available to them. Thus many species of vertebrates and some insects (Fitzpatrick and Wellington 1983) live within a territory or a home range. The difference between these two concepts of space use is that while a territory is defended against conspecifics, a home range is not defended and may be shared with others of the species simultaneously or sequentially. There are a few animals, e.g. some primates, which combine the two and have a large group territory, the fringes of which they may not range over regularly. The two kinds of biological space serve different functions. A territory provides the holder with priority to resources there, either food or mates. A home range is patrolled by the occupant(s) more or less regularly, and hence allows the animals to become familiar with local resources, and to deploy their activities to maximize returns. Such restricted space occupation means that the animals have a sensitivity to habitat configuration, and are responding to complex signals within the environment.

Some animals carry a mobile territory around them that is called personal space. Conspecifics are kept at a distance, unless for specific purposes of mating or grooming, etc. This is a social rather than a habitat concept, and is described in Chapter 4.

There is a broader scale of habitat choice for animals which live in situations where environmental factors may place physical or chemical limits or create optimal conditions, e.g. osmotic stresses in brackish water or illumination level, etc. Animals can move directly away from such limits or towards such optima by a mixture of reflex kineses and taxes (Fraenkel and Gunn 1961 and see Chapter 8).

There is another approach to considering space limitations on animals. It is to consider whether animals are scattered evenly, randomly or in clusters through their habitat. For animals which live in opaque habitats so that they can only be investigated by remote sampling, these spatial patterns are investigated by statistical analyses of quantitative collections taken, for example, by benthic grabs. The procedures have also been applied to animals which live over geographical scales beyond easy sampling by investigators, e.g. analysis of waterfowl and wildlife counts from strip transects. Evenly scattered animals have a mean population density much greater than the calculated standard deviation. In randomly scattered animals the mean density equals the standard deviation, and in clustered animals the mean density is less than the standard deviation (Elliott 1979). There are various statistical tests for estimating the significance of the differences, and hence the probability that the statistics indicate real spatial patterns. If the patterns are real, they indicate that the evenly spaced creatures have behavioral mechanisms or environmental forces keeping individuals apart, whereas clustered individuals have aggregating, possibly social, behaviors and randomly spaced individuals are not interacting with one another nor forced together. Most such statistical measures have shown that animals cluster, and even where individuals may be evenly or randomly spaced, they may be so only within broader scale clusters. The statistical approach is complex and limited, and needs direct behavioral confirmation by observation.

Some animals sequence their habitat requirements and migrate from one place to another. These movements can cover distances which are enormous relative to the size and locomotory ability of the animals. They may also be derived from or be assisted by physical displacement by wind or currents. Such long-distance movements entail a range of sensitivities to cues which are effective over long distances or which can be sequenced while en route. They may be substantially different from those cues effecting short-distance movements (see Chapters 7 and 8). Migrations of this kind occur among many of the active phyla including all vertebrate classes and many arthropods and cephalopods. Population crowding may also induce emigration or dispersal, drawing on the capability for long-distance travel which normally is deployed for periodic migration, e.g. lemmings can both migrate seasonally and emigrate during localized population explosions.

The habitat sequencing may vary on short time-scales, as with birds which roost in one part of their home range and feed in another. Or the sequence may occur over long time-scales as with the pelagic larvae of benthos which must transfer from the water column to seabed or shoreline, and have adaptive behaviors assisting them to do so, e.g. chemically and mechanically sensing the substrate.

In short, many animals change their habitat during their lifetime, either periodically or sequentially. Once in a habitat which is generally satisfactory, they have behaviors which can assist them in avoiding limiting conditions, or by which they can place themselves in optimal conditions. They may also have behaviors by which they restrict themselves to only a limited part of the space available to them. In this way they get the benefits of priority to resources, and familiarity with the habitat. Their mobility (vagility) may be substantially less than it appears.

Inactivity

Inactivity is important since it has a number of behavioral implications. Inactivity can be dangerous, particularly as it is often accompanied by the physiological phenomenon of lowered metabolism. There can be energy savings to an inactive animal if it can minimize its metabolic rate, or put conversely, when an animal can benefit from reducing its energy demands there are advantages in inactivity. However, lowered activity can mean slow escape responses and even reduced alertness and sensitivity. Accordingly, many animals which for one reason or another enter periods of inactivity, have associated behavior which puts the animal in the right habitat to minimize risks. Also, metabolic downtime may have overt behaviors such as reduced breathing rate.

Inactivity appears to serve at least four functions. They are fatigue, sleep, waiting and hibernation (dormancy).

Fatigue is the inactivity lasting a few seconds or minutes following anaerobic exertion. This includes the fast locomotion of escape or predation, and the fast action of aggression. In these activities normal aerobic respiration, in which carbohydrates are oxidized with the liberation of energy driving the behavior, is superseded by anaerobic respiration. Carbohydrates are incompletely oxidized to lactic acid, which accumulates in the muscles and eventually brings them to a stop until such time as delayed oxidation can eliminate the waste product. Thus the escaper must find its hiding place for protection before becoming exhausted, the ambush hunter must catch its prey quickly or seek another, and fighters must settle their battle before exhaustion renders them vulnerable to each other or to a predator.

Sleep is a term that has been applied in many ways to humans and animals. For animals sleep includes all those periods of inactivity with lowered metabolism which occur periodically, though not necessarily regularly, for short periods of a few minutes or hours. Thus humans sleep every day, usually at night, as do many other primates, mammals and other vertebrates. Some of these have adjusted their sleep to occur during daylight hours and are nocturnally active. Others such as some fish and birds may have two high activity periods, often at or near dawn and dusk, with two consequent lows: late at night or during the early afternoon hours. These often correspond to the lowest illumination period and the hottest part of the day respectively. The reduced metabolism appears to be an energy conservation adaptation (Meddis 1975) but may have other functions also. Traditionally it has been considered recuperative.

Sleep in humans is accompanied by characteristic electrical body rhythms as well as reduced metabolism and appears essential for most individuals. It may serve the function of processing information sensed during the preceding hours and placing it in a longer term data bank within the brain. It is often preceded by appropriate culturally determined activity leading to body insulation and such other phenomena as withdrawal from company or from illumination. It is accompanied by changes in posture, eyelid position, breathing rate, and periodic body and eye movements: REM (Rapid Eye Movement) sleep. Similar behaviors often occur in mammals and birds, and we can call them sleep without much argument. With increasing structural and physiological differences between other animals and the higher vertebrates it becomes increasingly difficult to apply the term sleep to short-term inactivity periods. The immobility of fish in an aquarium or river pool, for instance, has some behavioral similarity to human sleep in spite of structural and social differences and the lack of physiological evidence for lowered metabolism or characteristic electrical disturbances. But short-term immobility of invertebrates may not be sleep in the sense that we understand it.

Waiting is also usually a short-term activity. It is distinguished from fatigue and sleep since the animal retains its sensitivity. In fact, close observation often shows movable body sensors such as eyes and antennae actively deployed to maximize the input of available information. Waiting appears to serve the function of passing time, either by remaining still or by slow ranging while waiting or checking for appropriate environmental stimulii to which the animal is sensitized. Thus salmon will commonly wait hours or days in an estuary before ascending the river during a fall storm. They will also wait minutes in a fish ladder pool before leaping to the next.

In invertebrates little is known about the occurrence and functions of short-term inactivity periods. They may simply be reflex responses to reduced temperature and illumination, in which the benefit-cost trade-offs include reduced energy expenditure during periods of reduced feeding by visually hunting predators. However flatworms have been found "lying quietly under rocks, the muscles relaxed" in resting or "sleeping" condition (Jennings 1976).

Hibernation (dormancy) is a term applied to long-term inactivity periods such as where a mammal survives (usually seasonal) adverse conditions, when feeding is impossible or its net benefit is negative. Hibernation is commonly applied to winter inactivity spanning low temperature and other weather related stresses. Aestivation is the equivalent phenomenon for dry season inactivity. There are other analogous long-term inactive periods such as the diapause and freeze-resistance of insects. In short, there appear to be convergent adaptations to long-term inactivity, which at least in vertebrates is accompanied by a reduction in metabolism below the basal metabolic rate of sleeping. Such long-term dormancy is always preceded by behavior placing the animal in the right habitat for its long insensitive period. In vertebrates the adult or subadult places itself in the right habitat, whereas in arthropods the behavior may be for adults to place eggs for survival of the next generation or larvae to place themselves in appropriate habitat for survival in the next life cycle stage.

Attaining similar long-term metabolic downtime is a strategy adopted by lower invertebrates, bacteria and plants through the life stages of spores and cysts. Hibernation is a special expression among the warm-blooded mammals of this widespread long-term inactivity strategy.

Removing disturbing stimulii (comfort movements)

Many animals show behavior which serves the function of removing disturbing stimulii from the body surface. It is as though they itch. For some animals with flexible appendages, the removal behavior is sufficiently similar to our own that we can agree that they scratch. Thus vertebrates with legs or arms (mammals, birds, reptiles and amphibia) scratch on occasion.

However, there is more to removing disturbing stimulii than scratching, even for vertebrates with clawed legs able to reach finely defined itchy surfaces. Many of these will rub large parts of the body against a structure such as a tree. Others will induce grooming by conspecifics. In both cases, the behavior may have taken on a social function (see Chapter 4): marking territory or home range, or bonding members of a group respectively. Such cleaning may have been adapted as a symbiotic feeding strategy by cleaner fish and shrimps.

In fish, such itches can be removed by dashes to a firm structure, a quick rub, and return to normal water column position. These actions are often seen in aquaria and can at times be observed in salmon leaving a fish ladder or counting box.

Active invertebrates, especially arthropods, show a variety of actions in which one appendage is rubbed over another. House flies, spiders, and crustacea all show these appendage actions, which give the appearance of cleaning.

There is an internal extension of comfort movements involving behavior which flushes air or water firmly through near-surface body channels. Thus many vertebrates yawn, cough, or sneeze, or show other mouth opening and closing movements as though air or water is being flushed firmly or even violently through lungs or gills. These mouth actions (and associated operculum movements in gill bearers) range from little more than the slow regular breathing movements, through a deep yawn, to a violent sneeze (in which the whole head is jerked with the movement). In air breathers, the chest movements reveal the depth of air flushing, and in water breathers the speed and extent of operculum movements show the violence of water movement. In fish with open swim bladders gas may be released during or after these actions, and subsequent air intake behavior may follow as the fish restores its buoyancy.

Little is known about analogous behaviors which flush away disturbances from gills or other near-surface channels in invertebrates, although crustacea and echinoderms can flush gills and cloacas.

Comfort movements may be stereotyped, i.e. performed in a similar way each time, and even appear as Fixed Action Patterns. In some cases, as with duck preening, the behavior may have been ritualized and taken on a secondary or new primary function as a social display (threat, or bonding).

Perpetuating oneself

This category is the first of two (see Chapter 4 following) concerned with social behavior, i.e. behavior involving two or more individuals of the same species. Social behavior is essential for sexual reproduction. It is unlike the five types of behavior already described, which can often be undertaken by an individual animal on its own without responses to conspecifics. In fact, most simple organisms live solitary lives in this sense most of the time. They do not respond to conspecifics differently than to any other animal in their ecosystem. They may eat another conspecific if it is in the right size range for prey and does not respond by escape behavior. Such cannibal behavior can extend to the complex animals. Female spiders often eat males of their species instead of responding sexually to the male's courtship displays, or after mating if the mate does not escape quickly enough. Year old piscivorous juvenile coho salmon are ambush stream predators which will include among their prey days old dispersing fry of their own species.

Among living organisms, perpetuating oneself is achieved by two strategies. The strategy can be asexual, in which the animal or plant divides itself into two units or produces a smaller unit or units, each capable of growing into the adult form. Protozoa can split into equal halves or several units. Metazoa can bud off units, e.g. coelenterates. The strategy produces quick results and allows opportunistic population expansion under favorable ecosystem conditions. The drawback to the strategy is that the genetic constitution of each new unit is identical, thus providing no opportunity for the several forms of genetic recombination which can occur during sexual reproduction. The only way for new inherited capabilities to arise is by occasional and risky mutational change (see Chapter 5).

Genetic recombination allows repatterning of the DNA materials so that new properties can arise from the proximity of one DNA component to another. This is achieved in sexual reproduction by DNA reordering during meiotic cell division and by fusion of the DNA from two individuals with slightly different genetic constitutions. It is the fusion of gametes from two different individuals that imposes the necessity for social behavior at the time of perpetuating oneself. Almost all Metazoa use the strategy of sexual reproduction, most of them all the time. The advantages of new inheritable potential are great for animals inhabiting complex ecosystems with variable properties, and also for animals in ecosystems which provide non-overlapping optima for different environmental parameters.

Almost invariably sexually reproducing animals have converged on yet another set of strategies. The cell fusing stage (haploid gamete) is encompassed within and protected by the vegetative stage (diploid adult). One of these gametes is active, a behaving organism in its own right, and seeks the other gamete which is immobile or at least sedentary. The inactive gamete can be produced in small numbers and provided with a large food store, thus facilitating protection during early growth, whereas the active gamete can be produced in enormous numbers with minimal food stores. We call the active mobile form the male (sperm) and the inactive provisioned form the female (ovum). Commonly adult animals produce only gametes of one kind or another, and have evolved in ways complementing the adaptations of the gametes. Thus male adults may produce enormous numbers of sperm and be promiscuous or polygamous, thus passing on their genes through many females. Female adults may produce few ova, and select the most fit of competing males. There are a wide variety of complementary adaptations combined in many different ways in different species. There are also oddities. The small marine bird, the phalarope, reverses the more frequent vertebrate male-female roles in self perpetuation in that the males preempt maternal behavior by occupying the nests and protecting the eggs and young. Some oysters and shrimps change sex during the individual's lifetime, as do some female cleaner fish if their group male disappears. Sexual behavior and male/female roles can vary greatly with species.

Not only do males and females generally have different behaviors within this sexual reproductive strategy, they may also have different forms, effectively functioning as sex badges providing signals of their sex to conspecifics (see Figures A2.6 and A2.7 in Appendix 2). These sex badges and behaviors are superimposed on species badges, i.e. species-specific form, and behaviors providing signals which are more or less effective in preventing hybridized mating between sympatric species.

In spite of the enormous variability in sexual behavior between animals the processes can be categorized for the purpose of an ethogram. There are three base categories: courtship, mating, and parental behavior. They occur in sequence.

Courtship is the initial stage and consists of those behaviors which serve to bring the two animals together at the right time and place for gamete transmission. It may be a very simple process in those species which liberate gametes to water for external fertilization. In these cases all courtship must do is ensure that gametes of both sexes are liberated sufficiently close in space and time that they survive until the other is available. This can be achieved by release of pheromones, or mutual sensitivity to similar environmental stimulii. Sessile external fertilizers such as oysters must settle in clusters containing both sexes, so that both types of gametes are liberated physically close enough to meet.

Many species, however, pass the active gamete (sperm) to the other sex (female) either for immediate fertilization or for storage until later (in cases where the balance of adaptations precludes the female being in the right place to deposit eggs or young, or being sexually mature at the time of mating). Sperm transmission can involve very complex behavior, especially in non-social species. In these cases male and female must associate together in a way that does not occur at other times, with precision activities and timing such that reproductive organs are brought together in the right way and at the right time. For large animals such as whales and elephants this can require skillful navigation. For fierce carnivores it entails the suppression of feeding or escape behavior. During courtship there may also be selection of habitat facilitating mate selection, the mating act, and subsequent parental care or growth of the young. The combination of requirements may be complex, and has resulted in many improbable, flamboyant or subtle behaviors, ranging from the displays of highly colored birds to the female initiatives of the beaver. In general, it can be postulated that where unusual (even apparently maladaptive) behavior occurs between two members of a species, it is courtship behavior leading to mating. (The major alternative for odd behavior is threat between competing individuals - see Chapter 4.)

The mating or transmission action is often fast and of such short duration that it can easily be missed. Salmonids, for instance, drop sperm and eggs over a few seconds following a complex courtship which takes two or more hours. Few salmon biologists have actually seen the mating act, whereas the courtship is obvious to any visitor at a spawning stream. Alternatively, there are many variations so that mating can be prolonged. Large crustacea commonly mate while the female is in the soft-shelled post-molt stage, and sperm can be placed inside her. Dogs and wolves may remain united for many minutes after actual sperm transmission, apparently as an adaptation preventing other males serving a still receptive female. Mice and lions can achieve mating repeatedly during a single bout of breeding, again apparently monopolizing the female's time and ensuring fertilization by the male that invested the energy in bringing her to the mating act.

Parental behavior increases the survival of the next generation. It may vary from nonexistent to so extended that it includes care for not just the next generation but the second generation after that. In cases where juveniles must be taught or given opportunities to learn survival skills, the fitness of the original parent is not secure until the second generation of young is started on its learning stages, which are derived from the learned behavior of the first offspring generation. The original generation must succeed in passing on the property of teaching or providing learning opportunities for young. Survival of the first generation, while dependent, is not sufficient if the second generation is to survive in its turn. Grandparenting or group (kin) care may occur.

The simplest forms of parental care are ensuring that eggs or young are placed in the right habitat for their survival as independent organisms. The placing must be done either prior to egg laying, etc., or after hatching (birth). Shorebirds with precocial young have a relatively short period of parental care after the eggs are hatched. More complex parental care comes with the parent or parents tending the young as they develop survival skills. They may be assisted by protective structures such as eggs and nests, some of which are constructed by the parents behaving in adaptive ways prior to or during mating. The most complex behaviors are those already described in which parents provide opportunities for the young to learn, or even teach by example and imitation. In summary, perpetuating oneself involves a set of behaviors which facilitate asexual or sexual reproduction. The latter, involving two individuals, has required the appearance of many complex activities which for convenience of the ethogram can be sequenced as courtship, mating and parental care.

Summary

A behavioral investigation can use an ethogram, i.e. an inventory of behavior, as a catalogue and checklist. A comprehensive list of essential and potential behaviors can assist investigators with specific objectives to understand apparently irrelevant behaviors. An effective ethogram needs to be hierarchically arranged to encompass diverse behaviors. A first order categorization for such an ethogram is: obtaining materials and energy (largely feeding), avoiding becoming materials and energy for some other animal (largely escaping), maintaining the right habitat, inactivity, removing disturbing stimulii, social organization, and perpetuating oneself.

CHAPTER 4. SOCIAL ORGANIZATION: THE ETHOGRAM CONTEXT.

Social behavior is the interaction that goes on between two or more animals of the same species. This may happen for many reasons. There are many activities which can be done more efficiently in company or in partnership than alone. Sexual reproduction, of course, requires interaction with another animal. Once selection pressures had evolved sensory and coordinative systems that permitted the perception and recognition of another individual as a potential mate of the same species, the capability to recognize one's own species could be drawn upon for other activities. In this way feeding, hiding, comfort movements, sleeping and migrations have all drawn on social behavior in one species or another, and so increased either efficiency or survival (or reduced risks). An example of social behavior in grouped animals is for one or more to stand sentry duty, either not engaging in the activity of the rest of its recognized group or, by being located on the fringe of the group due to age or rank, most likely to see a predator first.

Similar interactive behaviors have developed between individuals of different species, but these are conventionally called symbiotic or mutualistic behaviors, and are discussed separately (see Section 4.6).

With social behavior comes an opportunity for social animals to remain in groups for periods of time or permanently. A problem then arises in that several or many conspecifics are actively engaged close together in a single type of behavior. The problem is overt intraspecific competition. When all are feeding, how do the individuals disperse themselves, and sort out the inevitable conflicts resulting when, for example, two see food simultaneously and move to take it? They must stay together for whatever survival value is operating, but they do come into conflict. They are involved in a typical adaptive balance where they must find ways to maximize the benefits of several interlinked adaptive strategies. The costs of group feeding, which must be balanced against the benefits, include sharing available food if it is limited, going short if you are not as strong or as fast as other members of your group, and getting wounded if you fight with a group member over a particular piece of food when you both want it. Any strategy which can minimize these costs can add to the balance favoring survival. The strategy that has developed in response to this delicate balance is social organization. It is the "law and order" of animal societies.

Organization among animals is mediated by several behavioral tactics. It results in the individuals within the group establishing a set of priorities for their various resources. If priorities can be set at a moment of intraspecific overt competition, one individual has the option of deferring to the other, and the competitive, potentially damaging incident, is de-escalated. A further feature of the adaptive balance is now apparent. Individuals with low priority have higher costs for their group benefits than others.

There are several other characteristics of this strategy, which has been adopted in many forms by many different species. The priority list is not perfect and not rigid. Individual members of the group come and go through departures or death, and may change their capabilities through growth or learning. The priority list changes inevitably, and at any one time may be in some flux through group members reestablishing their priorities. This can be either because the group reassembles after periodically dispersing, or because an individual attempts to raise its priority rating.

Priorities apply to several resources other than feeding rights. Any resource which can be in short supply, or which is limited through the constrained range of the grouping individuals, is liable to require group members to establish a priority system. Mates may be competed for, as may partners to groom, or space to construct shelters, or to feed, or to perform the ritual behaviors of courtship. In the last two cases, space for feeding and courtship, the priority for a resource has been shifted to a derived feature, space, as a means of achieving the basic resource of food or breeding opportunity.

Groups of animals may occur as an aggregation by chance (eddies in streams) or as a result of favorable environmental forces (wood lice in optimally humid cavities). Aggregations are not organized groups. The individuals happen to occur together, and each responds to the other simply as another non-dangerous, non-edible animal within the ecosystem. Organized social groups in contrast have some continuity, and the individuals recognize outsiders.

Tactics for the strategy of organization

Organization is mediated through one or another, or a mixture, of four tactics. They are ranking, territoriality, schooling and caste (division of labor).

Ranking

Ranking is behavior by which animals form a linear order of priority for action. Rank may be maintained by force in that an animal can actually fight to enforce its rank. However, much of the time rank is maintained by displays of threat, that is by a display such as the baring of fangs, or the raising of hair or feathers, signalling attack. Thus an advantage of ranking to grouped animals is avoiding unnecessary social conflicts and reducing energy expenditure when animals repeatedly meet. Ranking can be so structured that the simple approach of a higher ranker will cause the lower ranker to withdraw. Each member of a ranking group usually appears to recognize all members as individuals. Their ranks are remembered, and others deferred to or displaced according to the rank of who is interacting with whom at any one time. There are a few exceptions, as has been shown in one species of small bird, Harris' sparrow, which lives in temporary flocks in winter. This species carries its badge of rank with it (Rohwer and Rohwer 1978). In winter the higher the rank the darker the color, or at least the greater the likelihood of a bird attacking another with lighter plumage or deferring to one darker. Thus these birds, which will meet many other individuals during the winter through changing groups, need not remember each one, but can see relative ranks at first sight. Such a strategy could be open to cheating and social disorganization, which may explain why carrying badges of rank is so rare.

Ranking behavior was first described among domestic hens by Schjelderup-Ebbe (1935), but he only recognized what many farmers had known intuitively long before. He described many of the base characteristics including the linear order, one hen being dominant over a regular order of subordinates, with the lowest ranker being subordinate to all. However, with more than about ten hens in a flock, the rank order became complicated and could even have loops, apparently due to the mixing of responses during the establishment of relative ranks. Hen E could dominate F, F could dominate G, but by some chance G might dominate E. It should be noted that the order is essentially linear, and should not be called a hierarchy.

There are necessarily a whole series of complications or consequences. First, male and females may include each other in a ranking order, or may have separate orders within a group. Mixed sex orders may be topped by a big strong male, but not necessarily. A female hyena is consistently the dominant in that species. Ranking may be decided by alliances as in some baboon species, and thus individuals who are not individually dominant can control group action. Female macaques can determine which of the bigger males has priorities when they are around the females. Size and strength may be related to relative rank, which often favors males, but this is not necessarily so among individuals of the same age.

Young animals are initially immune from ranking, and infants of many species can play-attack adults without provoking dominance responses. However, juveniles in stable groups, such as those of many primates, learn relative strengths and assertiveness during their extensive play, and ranks arise which are in part affected by their mothers' ranks. Eventually the juveniles grow and take on rank within the adult structure. That rank usually reflects their play rank and inevitably also the relative rank of their mothers. In some species, and this is documented in baboons, juveniles leave their natal group and join another, taking on high rank in the process. It has been assumed that they had relatively low rank in their natal group, and thus the group change allowed raised rank, and the advantages which come from it.

There is distinct fitness value in high rank. DeFries and McLearn (1970) showed that in caged, genetically marked mice the dominant of three males sired 92% of litters born to the three females in each of the cages. There are frequent instances visible in the field and on film/videotape records of dominant male wolves, lions and other carnivores taking priority in feeding at a kill. High rankers determine group activities, and hence group survival may depend on the abilities of the strongest, fittest individual. High rank has responsibilities, and the qualities of the dominant consequences for the group.

There is a tendency to assume that high rank is a goal in itself, and that grouped animals will compete for rank, even though the intraspecific competition may be mediated more or less peacefully. This is not necessarily so. High and low rank can be considered distinct strategies, each of which has certain benefits and costs which can vary with group affluence. In affluent times, when food is plentiful, low rank is not a particularly risky strategy for feeding, and if the balance of adaptations reduces competition for mates by monogamy or tolerates promiscuity low rankers may also breed. In hard times the risks of low rank increase, but this can be ameliorated by several strategies. Thus male wild turkeys establish rank in fraternal pairs, the subordinate of the pair still having high inclusive fitness.

Low ranking chimpanzees groom high rankers and get protection from harassment. When high rankers disappear (which may be by choice on occasion, as in zebras, rather than being killed) upward movement may be quite orderly, with few individuals taking part. Subordinates also have tactics to increase opportunities to perpetuate themselves. Subordinate salmon males can dart in alongside the dominant male and shed sperm over eggs just dropped by the courted female into her gravel nest. Subordinate male elephants in *musth* can dominate others around estrus females by their temporary aggressiveness (Eisenberg et al., 1971). In groups of animals only one can be dominant at a time, so it is not surprising that subordinates in some species have evolved counter strategies to increase their fitness. Thus higher rank need not be a driving force among grouped animals.

In one species, chimpanzees, a rank change was brought about by learned behavior. Mike, a member of Jane Goodall's tribe, learned to associate the noise of clanging kerosene cans together with deference from group members. Mike realized the benefits, upgraded his displays by incorporating can clanging, and took over dominance of the group from Goliath. He was eventually displaced a few years later by a single confrontation with Humphrey. His rank tactic was a bluff, not supported by appropriate prowess.

Mike had superior learning ability, at least in this instance, compared to other members of his group, and so made a rank change up from a relatively low ranked subordinate. In other species, there are subordinates which may have superior qualities that do not contribute to their rank. Among African hunting dogs one animal may be faster and consistently bring down the prey. Among the Gengis pack (Van Lawick-Goodall 1970) Swift did this, but when Gengis the dominant disappeared, Swift and the dominant female Havoc each held top rank briefly before Rasputin took over for an extended period. The dominant is not necessarily either the leader of the group nor even the initiator. The lead zebra or baboon may be an experienced female or energetic young male and subject to the dominant's veto, or redirection. The dominant is also not necessarily the best able to learn. Mike the Gombe chimpanzee rose to dominance through his clever association of noise with deference, but Imo, a young female macaque who learned to wash her food free of sand, being young and female was a low status subordinate. Her discovery was copied but her rank was not elevated.

Although ranking is essentially an individual strategy, in species where subgroups form within groups as in matriarchal elephant families, the subgroups may show ranking orders among themselves. Jezebel and her immediate family could displace other families according to the

Douglas-Hamiltons' (1970) report. In the Gombe chimpanzee tribe, one subgroup split off, and was later harassed and at least some members killed by the remainder of the original group.

In summary, ranking is a behavior establishing priorities within mobile groups and maintaining them relatively peacefully. Each individual is free to move within the group's space, subject to constraints imposed by a higher ranker, which may be so terrifying that a low ranker can be very constrained. Higher rankers either inhibit the subordinate or direct its actions by controlling group activities. The space freedom is in contrast to the second manner of mediating priorities -- territoriality.

Spatial grouping (Territoriality and schooling)

There are two forms of spatial grouping which serve the function of social organization: territoriality and schooling.

Territoriality is the process by which individuals divide up a fixed space so that each has priority to resources within a particular patch. They may have to defend their space in order to confirm their rights. This may be overt aggression, which occurs in sea gull colonies, especially during spring arrivals, or ritualized threat displays that substitute for the aggression. Thus a sea gull owning a breeding territory may confront another by raising its wings and head, and being ready to stab down. Or it may displace into grass-pulling at the edges of the space which it will defend against intrusion from a transient or a neighbor. Male sticklebacks ritualize threats into head-standing at their teritory boundaries. A territory is often defended by the male, with the female being one of the resources, but this need not be so. Among beavers, the territory is female initiated and organized. Among Pacific salmon breeding territory may be female initiated, and defended against other females while males compete until one establishes dominance over other males, and becomes the consort, defending his resource of the female.

Schooling is the general term used for the free-ranging herds, flocks and pods, etc. of various animals, especially vertebrates. An overall term is needed to reflect this convergent behavior of forming closely packed simultaneously orienting groups under certain conditions. Schooling commonly occurs in air and water column animals, marine nekton and flying birds, where the group is in a relatively unstructured habitat, other than the presence of conspecifics. Commonly when under attack schools will close ranks, which appears to serve the function of confusing a predator. Whatever the survival value of schooling, it is a widespread phenomenon, and where it occurs imposes spatial constraints on individuals. It is noticeable that schooled animals show relatively little if any ranking behavior, or aggression to each other.

Territory and ranking can be combined or alternated as tactics for mediating social organization. This has the advantage that groups can continue to assemble closely for whatever benefits arise from group living, but within the group spaces can be occupied in which each individual or pair has top rights. The consequence is that resources can be evened out among the group.

Spatial dispersion for optimizing resource use has been adopted by many species to meet quite different environmental situations. In rivers where organic detrital food flows in proportion to current velocity but may be concentrated in eddies, the river bed may be carved into territories by resident salmon and trout. Each territory has a waiting station, where the fish derives an energy assist from a standing wave, and may have points of good feeding where standing turbulences hold detrital particles, or attract smaller prey species. On land, kob establish small, temporarily grouped, mating territories within their total home range. The concept of a defended territory may merge into that of a home range, as the latter, which may be patrolled by many individuals separately or in groups, is defended or at least at the fringes avoidance action is taken against other groups. Thus the resident pods of killer whales in coastal British Columbia associate with some neighboring groups sharing their range, but avoid more distant groups which may only overlap with them occasionally (Ford and Ford 1981).

Although territoriality is facilitated by the recognition of the neighbor as an individual, this capability is not essential. Territoriality appears in the lower vertebrates, fish, reptiles and amphibia, and is maintained by aggressive behavior towards all intruders. However, when territories are grouped and established, the frequency and violence of defense is reduced due to recognition of the neighboring territories, or territory holders.

Territoriality does occur in some invertebrate animals. It is widely known among insects (Fitzpatrick and Wellington 1983). Some limpets defend territories of algal gardens created by their grazing reducing the flora to an early successional state. Territories may occur among octopuses; at least dens are consistently occupied by some individuals. It can be difficult to demonstrate territoriality in invertebrate animals because of the phenomenon of group owned space. Several octupuses may share a set of dens. Among the colonial hymenoptera, group hives or nests may be defended. It appears also that territoriality is a property of active animals, vertebrates and arthropods, which have the ability to recognize configurations of space, sounds and smells, i.e. a sensitivity to landmarks, or the pheromones or calls of conspecifics. Territoriality is an organizational strategy requiring fairly advanced sensitivity and coordination.

There is a related phenomenon of "personal space". It is particularly important in schooling as the means of preventing animals from crowding so closely that they bump into one another. An animal can be considered to carry a territory around with it, which it will defend if a predator or conspecific comes too close. In addition, many animals have outer personal distances: the alert and flight distances. Some have no defended distance but may be contact animals, huddling together for warmth, reduction of heat or water loss, or for other functions.

Caste (Division of labor)

The fourth way in which organization within a group of animals is brought about is by caste. The group is organized to function as a unit since certain types of activity that normally are undertaken by each individual have been taken over by particular individuals within the group. The activities usually taken over for labor division are "obtaining materials and energy", "avoiding becoming materials and energy for some other animal", "comfort", and "perpetuating oneself".

The degree to which labor division has progressed varies with the group such that in some species individuals may be structurally specialized (soldier ants) with no option to take up the activities of other castes. At this stage the group is, of course, functioning as an obligate colony, even though the individuals are not physically incorporated together as with the colonies of corals and siphonophores.

The labor division may be incomplete in that individuals may tend to serve in one set of tasks, but still retain the capability of serving in others (ventilating, cleaning, etc.). Also, tasks may change through the individual's lifetime, as with worker bees which after metamorphosis function as nurses to the larvae before taking on feeding duty outside the hive. There are many colonial and semi-colonial species, and an enormous array of labor division within the termites and ants/bees/wasps in particular.

There are few animals other than insects which have taken up division of labor as a means of organizing groups. In even the densest "colonies" of breeding gulls and other marine birds, each individual is capable of undertaking all the activities essential to the species, within its sex and age group. Even the temporary "slug" of slime molds is composed of individuals which are each capable of all the activities needed when they live independently.

Signs of division of labor come in species with non-breeding individuals. In mammals non-breeders may take over support activities while a suckling mother is more or less immobilized. Cooperation of this sort is temporary, or at least facultative in the sense that the non-breeders or supporters can revert to the full range of activities if it is functional for them to do so. The African mole-rat is unique among mammals in evolving a caste society. It is based on a large reproductive female, several male studs, large guard individuals, and small laboring burrower/feeders. It is believed that the colonies are inbred kin groups with little genetic interchange.

A form of labor division occurs in some human groups where specific activities needed by the social culture are assigned to specific castes. The significant difference between castes in humans and animals is that individual humans still undertake the full range of survival (as opposed to cultural) activities, insofar as they are allowed to develop and express them by the cultural practices of the society. Outcast and Brahmin alike still eat, defecate, commute, sleep, scratch and mate.

The advantage of division of labor for organization comes in groups of vast numbers of small-sized individuals lacking a complex brain. The individual animals only have to perceive and respond to the caste of another group member (i.e. the range of activities shown by the caste member) in an encounter. This can be mediated by relatively simple chemical stimulii, or the configurational visual stimulii of a common shape (rather than the complex configurational visual stimulii which would be needed to identify one individual among millions). Thus labor division, which appears among group-living and semi-colonial insects, correlates with the evolution of enormous colonies of termites and Hymenoptera. In contrast, large vertebrates have usually adopted the alternative organizational strategy of spacing accompanied by individual recognition and ranking.

Communication

When animals are organized into groups, there is feedback to the ecosystem about group presence and activities. The presence of a group and the information about it can produce impact on other compartments of the ecosystem. Group activities can modify or add to available information within and about the ecosystem as received by other animals. Figure 1.5 models a behaving animal and shows it receiving information from the ecosystem through its sense organs. Ecosystem information arises from both inanimate and animate components, the latter including information about prey, predators and competitors. With the capability of active animals to have an influence on conspecifics either through competition or social cooperation, more information is injected into the ecosystem by the activities of the animals themselves, either incidentally or by adaptations serving the function of communicating information.

Social animals have drawn on several modes of injecting information into the ecosystem, and transmitting it by visually perceivable cues, or by sounds or smells, and occasionally by electrical signals. It can also be considered that the adaptation functions to manipulate a conspecific (Dawkin and Krebs 1978), i.e to attract it or to drive it away. Thus courtship and threat displays serve the function of manipulating a mate or competitor to the displayer's advantage, i.e. the displays are not altruistic or undirected news bulletins.

Visually transmitted information is mediated by badges and signals. Badges are structures which communicate information about the potential behavior of the wearer. They serve many different functions, ranging from the level of transmitting that the wearer is a member of a particular species to showing the imminent appearance of particular behaviors such as attack or mating.

Most animals have an external appearance in common with other members of their species (see Figures A2.4 - A2.7 in Appendix 2). There is an advantage for members of a species to look alike (also to smell alike or sound alike) since this provides information by which conspecifics can sense and respond to them species-specifically, e.g. by mating, grouping or cooperating. These structural species-specific configurations may be so pronounced that predators or prey can also recognize them and thereby selectively prey on or avoid them. When food is abundant, a predator can select prey that gives optimal returns for effort. When a particularly dangerous predator appears, a prey may take avoidance action sooner than if a less dangerous predator were nearby. Biologists also draw on these badges in establishing taxonomies and recognizing species in the field.

Badges may be transient and indicate behavioral states. Thus angry or amorous octopuses flash red and pale colors and dark stripes in various patterns. Juvenile anadromous salmonids resident in rivers and lakes have species-specific color patterns which pale during escape and become overlaid with a silvery smolt pattern during schooled emigration of the fish to seawater. Even mammals with highly species-specific plumage badges can vary their colors with age. Silver or light-colored hair indicates sexual maturity in male gorillas and sea otters, and advanced age in chimpanzees, dogs and humans.

Badges may be flashed by the behavior of the animal, either through its locomotion or movement of specific body parts. Thus some fleeing deer and antelope flash a white rump, providing a highly visible signal taken to mean danger by herd members. The brightly colored species-specific wing specula of female dabbling ducks are flashed during courtship, thus preserving the cryptic coloration at times when there is no need to attract a mate.

Badge flashing may in turn have been incorporated in a stereotyped movement, an FAP called a display, and thus converted into a component of a signal. A signal is a behavior which serves the function of communicating information about the transmitter to a receiving animal. In principle, signals need not involve badges, but they almost always do since badges increase the perceptibility of the signal. Even relatively inconspicuous animals such as the females of many bird and marine mammal species may show subtle color patterning that adds to the visibility of a mating display. Signals have become stereotyped, or ritualized, with extraordinary complexity in some breeding birds, mammals, fish, spiders and slugs. Each of the courting pair must progress through the informing display sequence subject to the risk of not completing or not surviving (due to attack by the other, or by predators).

Signals can involve sensory modalities other than vision. Sound signals have also evolved to extraordinary levels of complexity, ranging from the simple alarm calls of birds, which can be recognized by other species, to complex species-specific songs, with repeated elements. The relative complexities are shown by sound spectrographs (Figure 4.1). There is some indication that insect pheromones in colonies or between castes inside and outside the colony may also be pulsed in complex ways (Wilson 1970).

Sound signals have been raised to a level of complexity in human language by which fine information can be transmitted. There is considerable controversy over whether language and its implications for thought and reasoning are unique to humans or whether other animals show elements of the distinctive communication that we call language. Attempts to teach chimpanzees and gorillas to communicate vocally with humans have not progressed beyond a few simple sounds, but an alternative procedure of teaching has shown the extraordinary capability of chimpanzees to learn a repertoire of sign language, and deploy the signals in novel

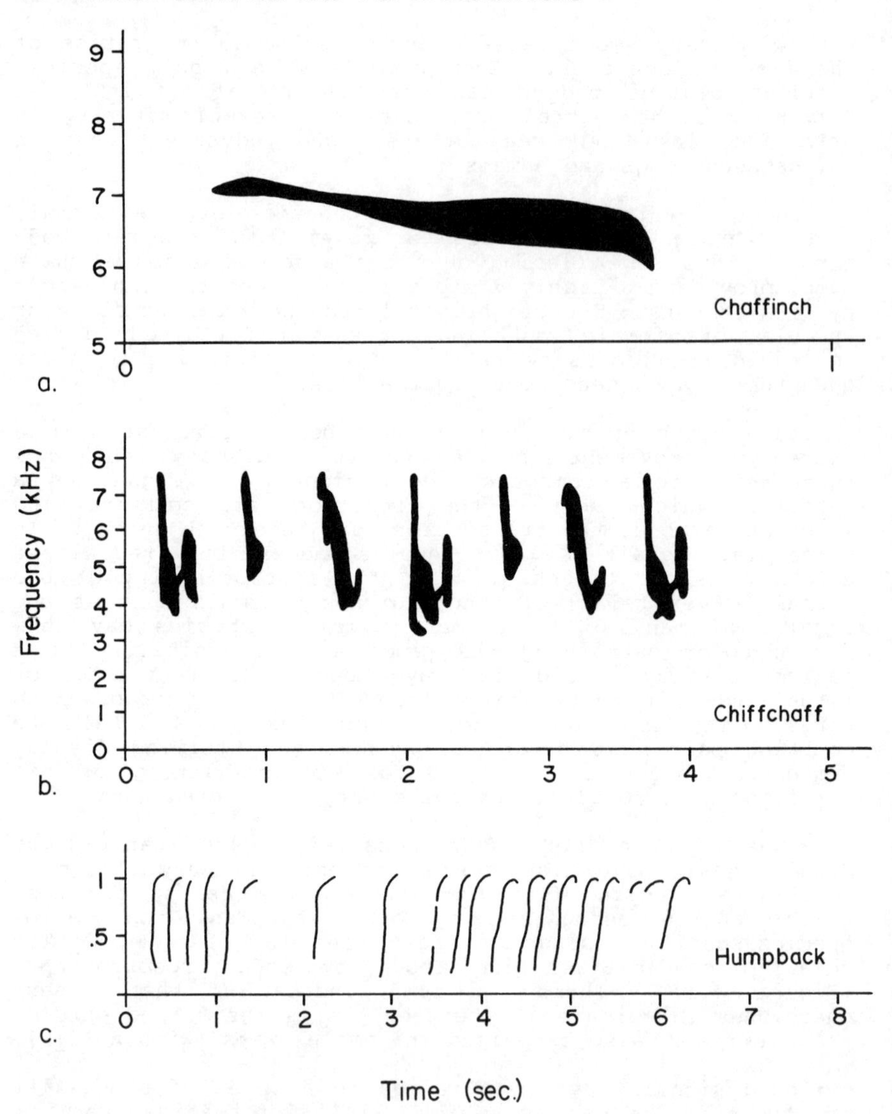

Figure 4.1. Sound spectrographs of: (a) a bird call which has the relatively simple element shown; (b) a bird song with complex elements and repetitions; and (c) part of a whale song (which extends over far longer intervals than bird songs). The horizontal scale shows time and the vertical scale frequency in kiloherz. Note that human sensitivity ranges from .02 to 20 Khz, so that we can hear these calls.

untaught ways (Terrace 1979) akin to reasoning based on language.

The production of chemical signals has diverged into two forms. A gaseous or liquid pheromone can be emitted into the ambient medium and function as a downstream marker. Receivers can direct themselves upstream if this is appropriate to the information provided about conspecifics and their sexual maturity. (Animals may also release chemicals by which predators or prey can sense them, and there will be selective pressures for acute olfactory sensitivity in this way and appropriate behavior for sensing odors flowing downstream.) Chemical signals may be secreted in dissolved form and available for habitat marking by rubbing, or they can be squirted for marking or for defense. Secreted chemicals retained on the body may communicate information to other conspecifics in close contact. Smelling body odors can be part of the behavior of courtship or social organization.

In a few species electrical signals may also be transmitted for courtship, aggregation and threats.

In summary, social species in many animal phyla have converged on the limited number of strategies for communicating information about the activities of individuals and manipulating conspecifics, often drawing upon visually mediated badges and displays, but also producing informative sounds and smells. Animals respond to the body language of conspecifics, competitors, and predators; and biologists can learn to recognize that language also.

Altruism

In animals which are not organized into groups, relationships between the individuals are competitive, the only exception being the limited cooperation between males and females for sexual reproduction. Competition arises either directly through simultaneous moves of more than one animal to take, for example, the same food, or indirectly where some of them have far greater individual fitness than others and can take the greater part of the food or other resource available in the area.

When survival results from group living, this simple dichotomy of competition and cooperation extends to being another of the many benefit-cost complexes that animals must balance in their interacting adaptations. An animal's individual fitness is extended to greater inclusive fitness if the activities of the group increase its survival and potential to perpetuate itself. It appears that many such groups are composed of kin, either closely related, e.g. brothers and sisters, grandparents, uncles, aunts and first cousins, or more distantly related. Under these conditions, the behavior of an individual which increases the survival of group members, even if it reduces the animal's own personal survival chances, will facilitate the perpetuation of kin genes to yet another generation.

It is noticeable that cooperative behavior is common among the individual members of groups, and that in many cases these groups are comprised of kin. Accordingly, it is important in observing the behavior of group living species to notice that there can be a range of activities not occurring in solitary living species, and also the kin relationships (if any) of the animals to each other.

A simple schema illustrating such behaviors is shown in

	Helps me	*Hurts me*
Helps another	Cooperation	Altruism
Hurts another	Competition	Spite

Figure 4.2. Matrix showing the social consequences of competitive, cooperative, altruistic behavior and spite.

Figure 4.2. Essentially, if my behavior helps me but hurts another I have won out in the competition, whether it be interspecific, or intraspecific within a group. But if my action helps both me and another, provided the other is also active towards the same end, we have cooperated. If a behavior of mine should reduce my chances of survival I am not likely to do it even if it benefits another group member. This would be altruism. It seems that it should not occur. Nevertheless, sentry bees do attack hive raiding wasps, and thereby die while the hive survives as a colony. Altruism occurs and each postulated case needs explanation in terms of how it could have evolved and its survival advantages (see Chapter 10). "Altruism" is a label for a phenomenon, and is not to be taken as an explanation. Put briefly, biological altruism may reduce individual fitness but increase inclusive fitness. It is not altruism in the sense that we use the word for human selflessness. There should be a payoff in that the chances of the family genes being perpetuated are increased, even though not by the altruistic individual, and this needs demonstration by a survival analysis. The group also may have increased survival but this is a consequence of the altruism and should not be taken as the cause (see Chapter 10).

There is a fourth compartment to the matrix in Figure 4.2. It is labelled "spite". Spite hurts both the behaving animal and others, and hence in theory cannot occur. In fact it may occur. I have seen a group of Oregon juncos around a feeder atacking any of their own group members that approached the food, so that none benefitted. Presumably it cannot occur frequently, nor for any length of time. Spite does of course occur among humans, but there it may have no genetic consequences.

Field behavioral studies currently include descriptions of altruistic behavior. It is included here as an ethogram component since such behaviors might otherwise not be noted, or might be discounted, unless the observer is aware of the concept and current theoretical interest in it. Nevertheless, the ethogram user must be aware that the term implies considerable interpretation of the function of whatever behaviors are labelled "altruistic". It is not a descriptive term of the type which is used to categorize "fighting", "feeding" and "mating" actions. When altruism is put into context with "competition", "cooperation" and "spite" it can be seen as an interpretative rather than descriptive level of categorization.

Intraspecies differential strategies

When predator and prey interact, it is easy to see that the interests of one are served against the interests of the other. If the predator is successful the prey loses; if the prey escapes the predator loses. The interests of socially interacting individuals within a species can also be in conflict. Thus the male of a species is charged with large numbers of easily produced gametes and it appears in general that his best strategy is to spread them as widely as possible by inseminating as many females as possible. Conversely, the female of a species invests a great deal of materials and energy in providing her ova with a food store, and it seems, in general, that she should add to her investment by careful selection of the fittest males to inseminate her, provide superior genes to her few offspring and in some cases to help in parental care.

In practice, each species balances the generalized theoretical sexual behavior with other adaptive demands. Thus there occur social-sexual arrangements ranging from once only monogamy (swans), through sequential monogamy if the mate dies (gibbons), to virtually complete amiable promiscuity (chimpanzees) and possessive harem ownership (elephant seals). In the latter there may be a corresponding range of cuckoldry, from the opportunistic to the compulsive, serving the needs for gene dispersal against otherwise enforced celibacy or monogamy. In some species sex roles are more or less reversed with female initiatives in courtship (phalaropes) and paternal brood raising (Siamese fighting fish). The interests of one sex are not necessarily well served by the activities of the other, and one sex may need to make substantial adjustments on the normal repertoire of behavior. In the context of the ethogram it is important for the investigator to bring into the repertoire of behavior, records of male-female interactions, particularly when these differ from expected norms.

There are three other sets of conflicting roles within group living species where the interests of one role player are not well served by the interests of the other. There are occasions when the survival or breeding of one or more parents is not well served by the presence of young. Thus under starvation conditions or if sublethal accidents happen to the young, they may be abandoned. In addition, in harem societies such as those of langurs or lions, a new male dominant may kill infants sired by his predecessor. The females will protect their existing investment (infants) as long as possible, but if they lose them the females quickly come into heat and are reinseminated. In these cases the strength and power of the adults wins out over the interests of the defenseless young. Conversely, in some species the life cycle is built around adult adaptations, favoring the survival of the next generation. Some species can place the next generation in the right habitat or provide food for it from their own bodies by dying. Food availability may be direct, as when infant spiders devour the mother's carcass, or indirect as when Pacific salmon carcasses decompose in the watershed and recycle nutrients to the ecosystem which supports the young.

A further conflict of individual interests within groups arises among siblings. When the balance of adaptations imposes competition as opposed to cooperation between siblings, the interests of the larger and more powerful, normally the older ones, will conflict with the interests of the weaker, normally the younger ones. Thus the last born, the weak or handicapped infants will suffer in the competition for food, parental grooming and other resources provided by the parent. The extreme is reached in some colonial insects where early born queens kill the others.

Interests of dominant and subordinate individuals, or high and low rankers, are also in conflict. Since subordination is an inevitable consequence of ranking among group members, adaptations producing appropriate tactics to maximize subordinates' limited resource access are to be expected. These would include the eagerness of low ranking chimpanzees to accompany and please the dominant or higher rankers by grooming, and subordinate male salmon quickly entering the nest and dropping sperm during the final mating act between dominant male and female. My students call these the sneaky sub strategies.

Some subordinate animals have skills superior to those of their dominants, and they may be important to group survival. The Gengis pack of African Hunting Dogs (Van Lawick-Goodall et al. 1970) had two such subordinates, Swift and Yellow Peril. Swift was unusually fast and generally the first of the pack to close in on prey. He struck at the tail. Yellow Peril had learned to attack the nose of prey. This, jointly with Swift at the tail, gave the Gengis pack an advantage over the conventional flight distance of zebras to hunting dogs. There was little opportunity for adaptive feedback of this superior hunting pack's ability during the lifetime of these two dogs, and the low likelihood of many groups of zebras being reattacked by the pack. So the pack fared well even when prey were scarce.

The generalization for ethogram use of the concept of intraspecies differential strategies is that the behavior of one individual within the group may not reflect the behavior of another. This is readily apparent in describing the behavior of male and female, but may not be immediately obvious in the behavior of parents and young in variable environmental and social circumstances, or the behavior of siblings according to relative strength, dominance and time of birth. There may be other intra-group differential strategies not yet recognized and awaiting description. The observer needs to be alert to the possibility. A good example of the unexpected occurs in some species of cleaner fish where groups comprise one (dominant) male and a harem of ranked females. If the dominant male is lost, the dominant female changes sex and takes over complete dominance. She can, however, be prevented or displaced by the arrival of a male which was living independently. Thus these species have three male strategies: be a dominant harem owner, be a solitary male, or be a dominant female until the opportunity arises to become a harem owner.

Interspecific associations

The alarm calls of many species are either similar to each other so that several cohabiting species may be alerted by the calls of any one, or a species can be programmed inherently or by learning to come alert at the alarm call of another species. Thus the alarm calls of crows at the approach of a bird watching human can still many of the small birds which he would otherwise see by their movement. Responses of one species to the activities of another occur to the extent that many forms of interspecies cohabitation

	Helps me	*Hurts me*
Helps the other	Mutualism	Not known
Neutral to the other	Commensalism	Not known
Hurts the other	Parasitism	Not known

Figure 4.3. Matrix of interspecific associational behaviors (symbiotic behaviors) arranged analogously to the intraspecific behaviors of Figure 4.2.

can arise (Figure 4.3).

The general term is symbiosis and implies that one species is part of the habitat of another and that both may have co-evolved together. There are several lower level terms comprising a hierarchy of symbiotic relationships. Commensalism refers to an interspecies association in which one species benefits, but the other (the host) is not disadvantaged. Thus pea crabs of several genera inhabit the shells of large bivalve molluscs of the genus *Tresus* and the burrows of tube dwelling polychaete worms. They get protection and waste food (commensalism) but some may take their host's food and damage tissues (parasitism). Many large marine animals have crustacea, e.g. copepods, living on their surfaces, which appear not to be parasitic in the sense of eating host tissues, but are simply living on an animate habitat. Relatively large examples are the barnacles which grow on whale torsos.

Mutualism is an interspecies association which is mutually beneficial. Thus cleaner fish and shrimps have a mutualistic relationship with a series of temporary hosts as larger ocean ranging and other fish come to a station occupied by the cleaners and are given a grooming which provides food for the cleaners. Decorator crabs can attach sea anemones to their shells. The crabs benefit from the stinging cells of the anemones deterring some of their predators, and the anemones are carried about close to a source of food, the wastage of their rather sloppy hosts.

The best known interspecies association is that of parasitism, in which the host is disadvantaged in the sense of providing food for the parasite. In extreme cases the host is killed, and if this is part of the normal life cycle of the parasite it must have the capability to transfer at the time of death or beforehand.

These interspecific associations may involve two species in behavior appropriate to their roles. Disadvantaged hosts such as caribou in warble fly range may panic and try to escape during the attack phase. Advantaged hosts (e.g. crabs) may actively seek and gather up their associates (e.g. sea anemones), which may relax and allow themselves to be manipulated. The benefitting smaller associate with a neutral or disadvantaged host, especially if it is larger and rarer, may have to seek a host in the way any specialized free-living animal has to seek its rare inanimate habitat.

In the context of ethogram use, observers should be prepared to look for interspecies associations and appropriate behavior encouraging or discouraging the association, depending on whether the species is advantaged or disadvantaged by it. The appropriate behaviors will correspond to ethogram categories of escape and habitat search.

Summary

Social organization is mediated by a number of behaviors which result in the reduction of intraspecific competition. These include ranking, territoriality and the division of labor. The inevitable feedback of information about an animal's activities to the ecosystem whence it can be perceived and responded to by other animals has been extended by convergent adaptations for transmitting information. Communication systems have developed through visual cues, sounds, smells, and occasionally electrical signals. When animals live in groups the converse of competition is cooperation. Cooperative and even altruistic behavior occur among members of groups. Such groups usually are composed of more or less closely related individuals (kin) sharing similar genes. Cooperative and altruistic behavior increase the inclusive fitness of individuals. Individual loss of fitness is offset by increased inclusive fitness, as kin genes are passed on to the next generation by surviving relatives. Individuals within a species may have conflicting interests such as those of male and female, parents and young ranking subordinates, and siblings. Species have acquired intraspecific differential strategies meeting the conflict of interest of individuals in these roles. Interspecific associations occur with individuals engaging in mutually responsive behavior analogous to social behavior and organization.

In the context of the ethogram, all these behaviors may occur and need to be noted.

CHAPTER 5. DEVELOPMENTAL CAUSES: HARD OR SOFT WIRED PROGRAMMING

There used to be a great deal of controversy over whether behavior is instinctive or learned, whether nature or nurture is more important in determining what an animal does and how it does it. The argument is too simple, too black and white. It is about whether behavior is genetically programmed, so that the animal is an automaton (simple or complex) with invariable behavior released according to stimulii received (hard wired), or whether the animal is even more complex and has some capability to modify its behavior according to stored information acquired through prior experience (soft wired). The wiring terms come from computer concepts. Furthermore, the differences between the concepts of inherited or learned, rigid or flexible, and one response or options, are not equatable. Rigid responses can be learned. A behavior can be reshaped with or without options.

The initial objective approach to this old controversy, as usual, should be inductive: to collate information showing the degree to which behaviors are pre-set, or conversely, are modified with experience. The results show how the controversy is a simplification of real situations. For example, the presence of complex chains of behavior in many species could mean that there are corresponding invariable chains of automatic responses. The approach of analyzing such chains has been followed. It has been shown, for instance, that stickleback breeding behavior consists of sequences of actions which are responses to relatively simple stimulii which function as signs for complex environmental and behavioral situations. Thus the egg-swollen belly of the female releases the zigzag dance of the male (Figure 1.2), and a sequence of other interactions leads on to egg laying and fertilization. But within each part of the total sequence, the animals have some allowance for modifying their behavior according to the local situation. Thus the male can direct his zigzag dance to the female whether she is waiting to the left or right of his nest. He can patrol his territory, concentrating on stretches where a neighbor usually appears. There are some components of the whole complex which are invariable, and others which are not. The mechanisms causing this mixture of the invariable and variable are difficult to unravel.

An alternative approach to understanding the significance of pre-set and modifiable components of behavior is to consider which behaviors of animals are functionally effective when first performed, and which change with performance. Within the first of these two categories there is a subset comprising behaviors which are performed only once during a lifetime so that if not performed effectively on that one pass, the animal does not survive or breed. One-shot behaviors are especially informative in this context of behavior programming since there is no opportunity for a backup repeat performance. The driving program must be right the first time it is used. The alternative subset comprises behaviors which recur unchanged and always functional. With the behaviors that change they may be improved either by maturation or learning so that the behavior becomes functional after initial non-functioning, or it may be modified to allow options to the behavior within a range of environmental situations.

Fully adaptive, first or only performance

In many species mating is a one-shot behavior adapted to various difficulties. There may be problems for very rare species for one sex to locate the other, in overcoming physical barriers to internal fertilization such as with the hard shells of crustacea, or in the ferocity of the other sex rendering courtship dangerous. The habitat requirement of egg laying may be different than the habitat requirement for the mating adults. Some species have also incorporated the energy bonus from parental carcasses into their life strategy so that the dead or dying bodies of one or both parents can be used as nourishment for the next generation. Obviously this can only happen once.

The black widow and many other web building spiders illustrate several features of the one-shot strategy. The female initiates the parental web and remains within it. The male seeks a female resident within a web, and undertakes an elaborate species-specific courtship of antennal signalling, body postures and web strumming. These courtship displays are interspersed with behaviors avoiding the female's attacks. In many species they include not only darting away, but also free falls with a trailing web line by which the male can haul himself back to the female's web. These patterns can be seen wherever large web building spiders set up house. When a large and small spider are seen together in such a web, a few minutes of observation can reveal the complexity of the movements. They can be a mixture of invariable and variable components as the male goes through its single attempt to court prior to being caught and eaten after mating. The displays must agree with inherent norms for the species, but can be interspersed with variable escape behaviors. Towards the consummation the actions must become progressively less variable so that at the time of mating the final approach brings male and female together, locking their genital organs together in a species-specific manner. This must be achieved effectively the only time it is performed, otherwise the male does not propagate itself. If there is ineffective performance, that particular individual and its genetic composition is eliminated from the species' breeding pool.

Other one-shot behaviors can be found in almost all phyla and range over many of the ethogram categories already described. Some female mosquitoes take a single blood meal as adults, and although their host location behavior may be directed variably by many different stimulii, the final action of mouth part probe, insertion and suck must be performed sequentially and effectively.

Many habitat transfers occur only once during a complex life cycle, as with the settling of the great array of pelagic larvae from water column to seabed or shoreline. Bivalves, gastropods, polychaetes, barnacles and many other larvae settle by means of a complex sequence of increasing negative buoyancy, sensitivity to odors and texture of the right adult habitat. Parasitic larvae may go through corresponding one-shot search patterns for an intermediate or final host.

There are noticeable differences in the scale of one-shot actions. In some cases the behavior falls within the concept of a single FAP. But mostly the uniqueness of the behavior is also at a higher level as a variable sequence precedes a final FAP. Both the final FAP and the overall sequence are one-shot behaviors, although the actions directing the animal to its goal may have been used repeatedly to direct the animal to other goals, and will be used again. Thus even one-shot behaviors can involve complex programming allowing sequences of low level variable and unique species-specific actions which are performed only once. Care needs to be taken in noting what the one-shot component is, and to what extent a high level one-shot behavior comprises lower level repeats and a one-shot FAP. Once these relationships are established we are in a better position to understand to what extent programming of the behavior might be determined in an invariable way (hard wired) or by means allowing alternative responses (soft wired).

This concept of variable and repeatable behaviors leading to an act terminating the sequence (which may occur only once in the individual's lifetime) fits within the concept of appetitive behavior and consummatory act (Craig 1918).

The concept directs physiologists to search for mechanisms allowing a stereotyped terminal act to be preceded by variable behaviors placing the animal in the right environmental situation to receive the stimulii for the consummatory act. Thereafter the behavior may lapse (if one-shot) or may be repeated sooner or later.

There are other species which repeat actions that are performed effectively on both first and later occasions. Pacific salmon will spawn about ten times within a period of a few days while they are on the spawning gravel. Each spawning, whether by male or female, is fully effective whether performed only once or the ten to twelve times until all gametes have been discharged. In both sexes there are potential disruptions to the fish. Either may be displaced by a more aggressive intruder, but more likely are displaced physically by storms and freshets which change current velocity, depths and turbulence patterns so that their prime nest building site is less satisfactory or even impossible temporarily.

In each case there is a competitive advantage for the first performance to be functional. It just so happens that there are often repeat opportunities. These may not occur, or the individual may be in perpetual competition with others of its species, as with food begging when among siblings.

Maturing behaviors

Some behaviors mature with their owners so that as time goes on they are performed more effectively. The behaviors keep the same essential function. There simply is not the need for early performances of these behaviors to be fully functional. Also, there may be no change in the association of stimulii and response, merely improved performance based on improved structures to perform the action.

Examples of behavior modification come largely from animals having a long development during which the young are protected from their inadequacy by parents, or put conversely, when a long period of parental care allows the young to complete their structural growth without immediate efficient performance of their behaviors. There is often considerable learning going on simultaneously, since long development allows that adaptation to proceed as well. It may be difficult to distinguish the processes of learning and maturation. A number of well-known cases in the literature illustrate the maturational and learning effects which can occur.

E. H. Hess demonstrated some differences in domestic chickens' abilities to peck accurately and precisely. He glued down grain in front of test chicks on an impressionable surface and demonstrated pecking ability by

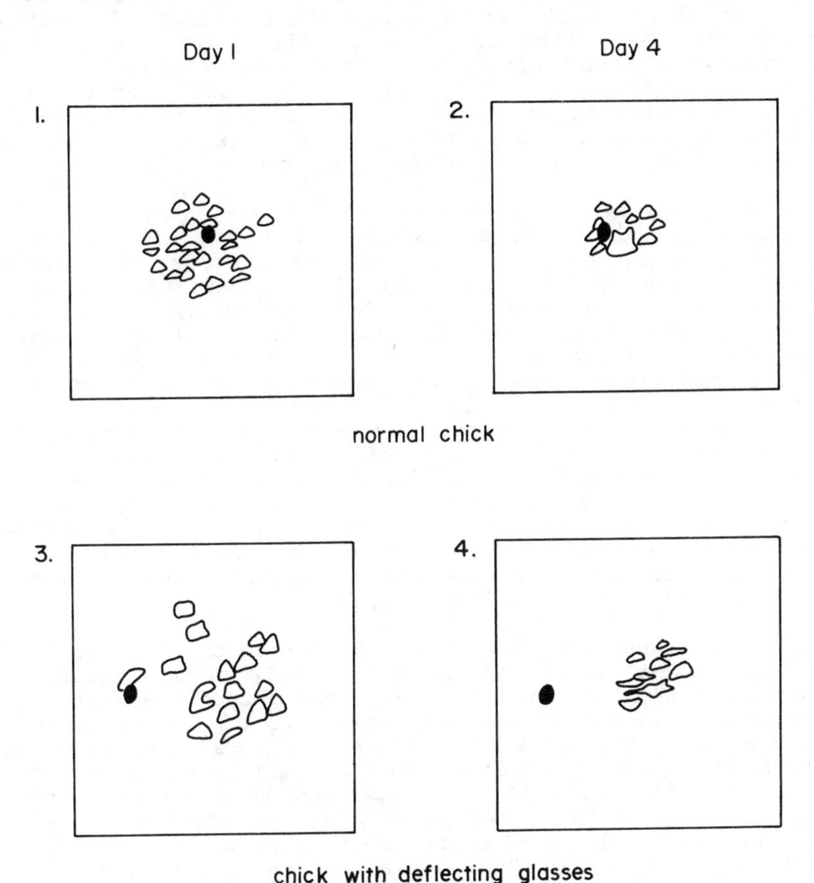

Figure 5.1. Chick pecking -- improved precision with age but inability to learn greater accuracy. Black spot - seed glued to surface; open spots - peck marks.

recording the position of peck marks (Figure 5.1). Very young chicks could peck in the general area of the grain (they were accurate) but their pecks were usually misplaced randomly all around the target grain (they were imprecise).

As they grew, precision increased even though they were not rewarded, since the grain was not eatable. Finally they hit the grain at almost every peck. However, when Hess fitted image shifting glasses to his chicks, he could make them experimentally inaccurate, with their pecks deflected consistently away from the area of the grain. Precision improved with normal growth even though the chicks remained inaccurate. Within the period of the experiment, limitations within the brain of the chicks prevented them from modifying their behavior to peck accurately, whereas the precision of their performance matured.

Other examples come from bird development, as is shown by Hailman's (1969) detailed analysis of a sea gull chick's improving performance in the nest and outside it (Table 5.1). Pecking at the bill tip of a parent becomes more accurate, and is eventually rewarded with regurgitated food, after which it becomes very accurate. Later the chick pecks indiscriminately at objects and marks on the ground, but quickly modifies behavior to peck at and pick up for swallowing only those objects which are food. Later still, as it learns to fly, it goes through a series of landings in unlikely places such as bush canopies until finally it lands only in firm places which will support it. It also narrows down the range of situations in which it tries to drink and bathe. At least some of these actions have been shown by experimental analysis to consist in part of learning in the sense of associations between stimulii and rewards reinforcing effective performance.

Other actions also change with performance. The sea gull chick initiates many actions incompletely or inadequately. It attempts to stand, to preen, to walk out of the nest, and particularly to fly. The wing movements of the latter may be initiated days or weeks before the total movements of flying can be performed effectively. Experiments with other birds have shown that if the chicks are constrained so that they cannot perform incipient flying movements, they nevertheless still fly effectively at the same time as unrestrained peers. They appear to be trying out their motor organs as they become movable, even though it is not necessary for them to do so.

Play in young animals, particularly young mammals, appears to be a mixture of these two processes of maturation and learning. It provides the opportunity to perform actions which will be functional later, but under circumstances where effective performance is unnecessary. It occurs at a time when motor structures are developing, but the sensors and brain are competent to acquire and remember enormous amounts of information relevant to releasing and directing the actions later in life when they must be functional.

Table 5.1. Sequence of actions on hatching of a sea gull chick (modified from Hailman 1969).

1. Initial act - stretches neck, pushes off eggshell lid.
   - a special muscle involved in neck thereafter shrinks and vestigializes.
2. Lies quietly. Brooded by parent.
3. Commences pecking movements at parent's bill tip when bill accidentally or deliberately put near the chick.
4. Pecking more frequent and accurate. Parent eventually responds by regurgitating food. Chick responds by swallowing.
5. Chick attempts to stand. Finally stands.
6. Chick preens.
7. Chick walks.
8. Chick walks out of nest (about 1 day old).
9. Escape behavior now consists of freeze response to parent leaving while sounding alarm call.
10. Escape behavior modifies to - run out of nest, enter cover and freeze.
11. Escape further modifies to - run to constant hiding place and freeze.
12. Start flying movement (1 week).
13. Start calling at strange birds (2 weeks).
14. Attack strange birds.
15. Feed independently. Peck indiscriminately.
16. Peck only at food.
17. Fly. Land clumsily at first, downwind, on tree leaves, etc.
18. Drink from any glittering surface.
19. Drink from water.
20. Bathe - anywhere.
21. Bathe in water.

Modifiable behavior

Much behavior is repeatable, i.e. performed more than once, particularly in the more complex animals. The opportunity for repetition increases with the increasing size and extended lifetime of the animal. In general a large slow growing animal has to feed more times during its growth to large size than a smaller animal to its small size, and unless it is a top predator the large animal will more frequently experience dangerous situations to which it must respond in order to stay alive. Accordingly, it needs physiological mechanisms which can be drawn upon again and again, or put another way, it must be programmed so that functional behavior can be released quickly and effectively whenever needed. This programming concept raises a number of issues, such as to what extent the program produces "invariable behavior", or "behavior which is modifiable to minor or major variations in environmental and social situations". Adaptability to behavioral modification can be grouped into two contexts: modification with improvement, and modification with options.

Modification with functional improvement

The everyday word for behavior modification following experience is "learning". We experience it ourselves and we can see it in the mammals and birds that we regularly encounter in our daily activities whether they be domestic pets, farmyard stock, hunted game or garden wildlife. All these animals learn where and when we will provide them with food, deliberately or unintentionally. They can learn when to avoid us or when (for dogs and cats) comforts such as affection and grooming may be available.

Similar associations have been demonstrated in such a variety of animals that we can expect the capability to occur at any level of complexity. The literature on the subject is enormous. For a general introduction see any introductory psychology text; but we need to be introduced to it in some way relevant to the biologist as opposed to the physiologist and psychologist. The biologist's questions should be of the type: "in what way is learning adaptive and contributes to fitness?" and "are there different kinds of learning with different survival potential?" Such questions can lead the biologist as far as he or she wishes to go towards understanding the ultimate or proximal causes of the learning, prior to black boxing the topic.

The first understanding (which has stood the test of time) of a type of learning came with the studies of Pavlov into what is now termed classical conditioning. His paradigm relating conditioned and unconditioned stimulii and

Classical Conditioning

1. US ⟶ R   2. US ⟶ R   3. US ⟶ R   4. US ⟶ R
S2   CS   CS ⟶ R

Operant Conditioning (trial-and-error learning)

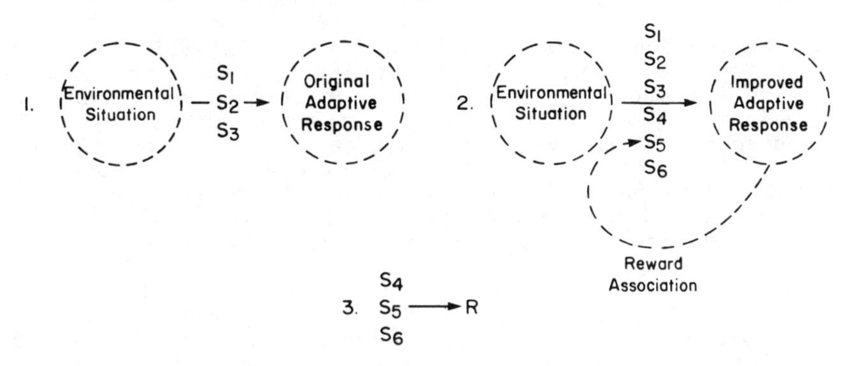

Imprinting

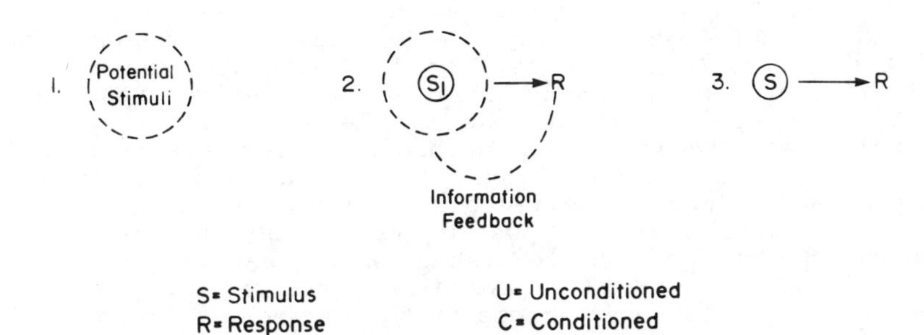

Figure 5.2. Learning paradigms.

response is modelled in Figure 5.2. Where a response can be shown to be released by a particular stimulus such as when a hungry dog salivates at the sight of meat, then that stimulus can be considered unconditioned for the purpose of the experiment, and a new stimulus can be conditioned to produce the same response. Thus each time the hungry dog is shown meat, a bell can be rung. After a number of trials with both stimulii together, the hungry dog can be tested with the bell alone. If it salivates, then the sound of the bell has become a conditioned stimulus. There are obviously a great number of conceptual and experimental refinements of the basic paradigm. These may explore which stimulii within the sensitivity of the species being tested can be associated together, the time and space relationships of the various stimulii during conditioning, and the presentation of "clean" stimulii in order to isolate the real stimulus that has been conditioned. An example of a "dirty" experiment is the dog being conditioned to the experimenter reaching for the bell rather than ringing it.

In the context of adaptiveness, animals will condition in the Pavlovian form if they repeatedly encounter coincident stimulii, one of which is not conditioned to a beneficial response, and one of which is already conditioned, or is inherently effective. In practice much learning appears to be of another form. Animals associate new unconditioned stimulii with rewards or the absence of punishments (such as injuries) through processes of trial and error. New foods and predators come to be recognized, for example.

A controlled form of trial and error learning was developed by Skinner and is now termed operant conditioning. The paradigm is modelled in Figure 5.2. Through rewarding (reinforcing) an animal as it performs an action associated with one of a medley of unconditioned stimulii, one particular stimulus can be conditioned to produce the response. The conditioning process may take many trials depending on the sensitivity of the animal to the conditions of the experiment, its ability to associate reward with stimulii, and the degree of complexity of the response which is being tested. The process can be automated to provide stimulii and reward in a mechanically or electronically operated "Skinner" box (Figure 5.3), and responses can be shaped into almost unbelievable behaviors such as pigeons playing table tennis or pushing model baby buggies. The reward can be negative (punishment) so that the animal is conditioned to show aversive responses. Operant conditioning has been brought to extreme refinement by animal trainers both consciously and unconsciously (Breland and Breland 1966) and are significant in the management of both pets and domestic animals.

The Skinner box technique and operant conditioning procedures have been developed for even such simple animals as Planarian flatworms. These lack coordinating centers more complex that paired cerebral ganglia. Their lack of a blood system renders them particularly suitable for a line of research which has shown that vivisected individuals with and without "heads" retain learned information after

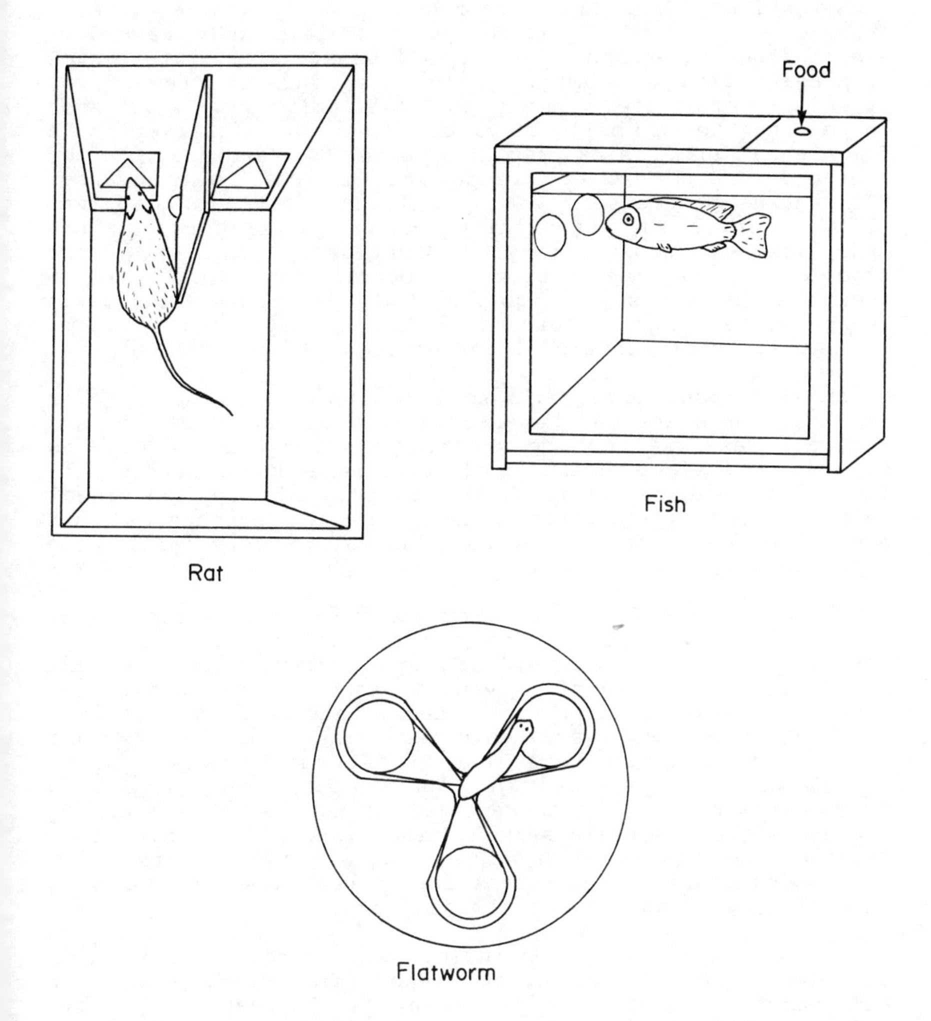

Figure 5.3. Skinner boxes for a variety of species and behaviors.

regrowth, and that learning can be passed to untrained cannibal Planarias. There appears to be some chemical basis for memory.

There is also a converse to learning which in humans is called "forgetting". A general term for it is "habituation", but like learning it may involve several types of processes. An animal once conditioned by reward or association to respond to a stimulus, will later stop responding if the stimulus fails to produce the reward, or stimulii dissociate. Thus the salivating dog will stop responding to a bell if meat does not follow, or a frightened gull chick may habituate to harmless shadows overhead and no longer freeze when a leaf falls nearby. Eliminating unnecessary energy and time costs is adaptive. At this level of potentially interacting processes, the underlying neurophysiology is obviously complex. For example, it may be easier to retrain an animal to do something previously forgotten than to train it to do something new, i.e. there may be a residual memory of some sort in spite of functional forgetting.

The important components to learning in behavioral terms are that responses can be associated with new stimulii, and that new responses can be adaptively molded to the demands of a new environment. Both of these consequences are functional and can add to the fitness of the individual animal. They occur very extensively in vertebrates, but also in some other active animals, especially the cephalopods.

There is another form of learning which was recognized by Lorenz (1937) and termed "imprinting" in English. This is generally accepted as a form of rapid, irreversible learning often occurring in early juvenile stages. The traditional case, for example, consists of newly hatched ducks or newly born calves rapidly learning to recognize their mothers, and thereafter following them instead of other female adults. The paradigm is provided in Figure 5.2. There have been some extensions of the concept to include juvenile learning of home stream odors in salmon. There are also complexities arising from misplaced imprinting on strange objects which may determine later adult behavior of the animal in choosing an obviously wrong mate or social group.

The adaptability of imprinting is in circumstances where an individual needs quick recognition of a complex environmental configuration such as "mother" or "home stream" either through dangers to itself in slow recognition of the only adult which will protect it, or because the individual may not stay long in the right habitat for its own later breeding.

Modification with options

Some behaviors require individuals to make choices. Thus predators may need to choose every day, where they will hunt within a home range which is not all accessible to them on any one hunting foray. While actually hunting, there are decisions to be made where to ambush, when to strike, when to abandon a chase. Some animals may be perceptive of complex details of their ecosystems. If their food species is patchy, speedy, armed with a battery of defensive behaviors deploying weapons, has extreme sensitivity to predators movement's and smells, or has the intelligence to perceive or remember ambush sites, then the predator may have corresponding optional behaviors.

We call animals showing such choices "intelligent". They have the ability to learn many features of their ecosystem, to retrieve learned information and coordinate it with their automated inherent and conditioned behaviors. There may even be occasions when such intelligent animals might appear to be reasoning, as though the choice of options is not immediately apparent. Dog owners have no trouble seeing such incidents. Lorenz (1955) describes how two dogs regularly meeting at a fence, and going through the motions of attacking each other, on one occasion found part of their fence had been removed. After a second or so of a dramatic pause, they mutually raced back to where the fence was intact and kept up their noisy threats to each other. Lorenz elsewhere (1952) describes how one of his mouth-breeding fish, when hungry but with a mouthful of young, was suddenly presented with food. Again after a pause (as though it was thinking), it carefully spat out its young, ate the food, and then retrieved the young without swallowing a single one. He wrote that the group of students with him were so impressed that they broke into applause.

The concept of learning, which can be defined operationally, grades into the concepts of "intelligence", "thought", "reasoning", "consciousness", "self-consciousness", and so on. The topic is vast, and for most behaviorists can be conveniently "black boxed" so that the concepts are acknowledged as potentials to be explored if apparently relevant to the adaptiveness of a particular species.

Hard and soft wiring

A calculator is both hard and soft wired. It is hard wired for simple functions like addition, subtraction, multiplication and division. A cheap calculator may be hard wired for squaring, so that once the number to be squared is input, only the squaring key needs pressing. More likely it is soft wired, so that the user squares a number by multiplying it by itself input a second time. Some calculators are hard wired for square roots, so that the number to be square rooted can be input and the single function key pressed, whereas in others the calculator is soft wired so that as with any other complicated function, a sequence of component simpler functions must be performed. In a programmable calculator, there is a driving hard wired program which accepts instructions in the form of an inserted program. The capacity to receive information and operate an inserted program is the soft wiring. It allows action according to the needs of the user.

In order to avoid the simplistic two-sided controversy about the relative importance of instinct and learning in the proximal causation of behavior, it can be useful to apply computer concepts of hard and soft wiring to the animal body, its inheritance, development and operations.

A hard wired animal is one in which nerve circuitry is set so that it responds in inevitable ways and times to stimulii. A reflex action or a Fixed Action Pattern is hard wired. It is conceivable that a totally hard wired life cycle strategy might be functional and the best possible if the animal remained within predictable ecosystem conditions. Such hard wiring could be determined genetically and hence essential behaviors appear fully functional when first performed -- at birth, hatching, or after metamorphosis. It is more likely that only some functions, such as the swallowing of sea gull chicks, benefit from being hard wired and other actions do not.

A soft wired animal is one in which nerve circuitry is not completely set, i.e. there are optional or initially incomplete circuits, so that outcomes of individual experience and growth can be input to determine subsequent development and operations of the wiring. Thus as the neck muscles and nerves of a chick develop and the reward of food comes from bill pecking, so directed pecking can be shaped as an effective feeding behavior. There are obvious potential survival advantages to such soft wiring. The animal can adjust its personal behavior to a range of ecosystem conditions, thus allowing it to inhabit conditions not requiring specialist adaptations, or alternatively acquire several specializations, each of which is functional in particular conditions. The learned behavior may become as automatic as a genetically hard wired behavior, but since in many cases the potential for further learning and behavior modification remains, it is to our advantage to remember the soft wired alterable circuitry inside. Wherever learning occurs or the potential for learning remains, we can consider the nerve circuitry as soft wired.

The matter of ensuring species-specific response to badges or signals of conspecifics can be achieved by hard wiring responses to sign stimulii representing the complex overall condition, such as red flashes on parents' bill tips. Alternatively the problems can be solved by common experience hardening up initially modifiable circuitry. Common experience is inevitable in species where there are structural constraints such as several weeks' growth inside an egg habitat. Kuo (1967) has identified common experiences which occur in such a habitat, and which may contribute to shaping later behavior.

In theory it seems that hard wiring would be very limited in some animals. It could be constrained to a few survival and competitive behaviors, particularly in early free-living periods, plus the provision of driving programs allowing the input of information from individual experience and growth. Some animals seem to benefit more from either hard or soft wiring than from the other. Short-lived species with complex life cycles would appear to benefit from hard wiring. There is little time for gaining experience, and the short life cycle is broken into separate sections, each with its own ecosystem adaptations. In addition, small species have less space for a coordinating system capable of storing information gained by experience. Specialist species, particularly where the habitat is rare or fleeting, may also be best powered by invariable responses, especially if they can breed in enormous numbers so that heavy losses can be tolerated. Table 5.2 summarizes patterns of adaptation which should be favored by hard and soft wiring respectively.

Some species appear to benefit from soft wiring. Long-lived species have opportunities for repeat experiences from which information can be stored and reutilized. Simple life cycles permit the carry-over of stored information throughout the lifetime. Large animals have greater spaces for brains. Tolerant species can range through a great variety of environmental conditions and may need different behavior strategies in various segments of an ecosystem gradient, due to the presence of various food species, predators or competitors in ecosystems with complex and variable compartments.

In this context, the original soft wiring allowing maturation and modification may harden with age or experience so that options are reduced or eliminated. Thus old mammals get set in their ways, and even gull chicks lose the potential to freeze from harmless stimulii. Their wiring gets hardened, which is a major difference from computer wiring.

Behavior genetics

The concepts of hard and soft wiring must be put into the context of what is inherited, and how the inherited material determines behavior. It is very noticeable that behavior in

Table 5.2. Life cycle strategies favoring hard or soft wiring of behavior (invariable or modifiable behavior).

Favoring hard wiring (Invariable behavior)	**Favoring soft wiring** (Modifiable behavior)
Short-lived	Long-lived
Complex life cycle	Simple life cycle
Small body	Large body
Specialist	Generalist
Rare habitat	Widespread habitat
Temporary habitat	Continuing habitat
Many offspring	Few offspring
High mortality tolerable	High mortality not tolerable
Small brain	Large brain
No parental adaptations	Parental adaptations

complex animals is not often inherited in simple Mendelian patterns. Convulsive seizures of mice may be so inherited (Collins and Fuller 1968) but this is not functional behavior; it is pathological.

At the functional level of behaviors, bird distress calls have been analyzed by the Mendelian technique of hybridization. McGrath et al (1962) showed that pheasant and chicken distress calls were inherited with the pheasant's shorter call dominant over chickens, but with intermediate frequencies due to the secondary side effects of size and anatomy. Dilger (1964) showed that hybridized lovebirds (*Agapornis roseicollis* x *A. fischeri*) with different nest material carrying techniques try ineffectively to carry like *roseicollis*, and end up carrying like *fischeri*. Instead of tucking several strips of material into their tail and rump feathers, they revert to single-strip carrying in their beaks.

In many cases of interspecies crosses, the hybrids are sterile, so that another traditional Mendelian analytical technique -- backcrossing -- is impossible. However, backcrossing can be performed within strains (or subspecies). The so-called killer African bee has shown genetic dominance in aggressive behavior when backcrossed with Italian domestic bees (Gould 1982). The measures involved the time elapsed before the first sting following a kick on the side of the hive, the number of stings delivered in 60 seconds on a test ball bobbed up and down 5 cm in front of the colony, the distance from the hive at which the last sting is delivered to the retreating ball bobber, and the number of stings on the ball bobber's glove. The ingenuity of the experimenters in devising a quantitative measure for genetic analysis was unfortunately not matched by their consistency in application. The tests of the African bees tended to be shortened from the intended 60 second ball bobbing period, and the retreat process by the testers speeded up. Italian colonies would deliver the last sting at a distance of about 20 meters, whereas African colonies would keep up the attacks regularly for 1000 meters on the fleeing testers!

There appeared to be a four-gene system determining differences in these measured qualities between the two races, but yet other genes involved in the determination of aggressiveness per se. Mendelian techniques demonstrate the genetic determination of differences in behavior, rather than the inheritance of functional categories of behavior.

Selective breeding within stocks provides the same sort of results. This has been the procedure followed pragmatically by breeders of domesticated animals. Gradually many breeds with differences more or less true to their strain have separated out. Some of these differences, such as the hygienic behavior of bees resistant to foulbrood, may be reduced to one or two genes (Rothenbuhler 1964) such as those controlling cell uncapping and removing pupal corpses. Such information permits understanding strain purity, and directing breeding; thus it has pragmatic value. However, it does not penetrate to the level of understanding behavioral causation in genetic terms, or how genes determine behavior by chemical processes during ontogeny.

Progress so far in such understanding comes from analyzing the single gene determination of differences in behavior, rather than gene determination of the functional behaviors of the biologist.

Genetic dissection of behavior ranges over organisms from bacteria to mammals. The simpler invertebrates show phenomena not immediately relevant to more complex animals, such as induced reduction of polyploidy in the ciliate *Paramecium* to haploidy and chemical treatments producing forward, fast and "paranoid" (membrane repolarization deficit) behaviors. The fruit fly *Drosophila*, as might be expected from its use in traditional genetics, has been productive in the genetic analysis of proximal behavior causation at the molecular level. Many single gene behavioral differences have been shown, and attention directed to the control mechanisms.

We understand reasonably well the patterns by which some structures and physiological mechanisms are inherited and determined biochemically as a result of coded instructions within the DNA. Behavior, however, is the product of the integrated interaction of these structures and physiology determined by the ultimate pressures of natural selection. Behavior is so complex, affected by so many internal physiological processes during manifestation and ontogeny, that we should not expect the biologist's unit behaviors to be determined in the simple Mendelian ways of dominance and recession. We should expect even rigid stereotyped FAPs to be the product of natural selection refining complex interactions of structures and physiology, each determined by the interaction of many different genes. Behaviors are essentially polygenic, and in complex animals functional behaviors will usually have to be broken down into many component parts before we attempt to ascribe parts to different genetic controls. Accordingly, the topic of behavior genetics is concerned largely with analyzing genetic controls on differences in behavior between species, strains and individuals. With this approach it may be possible eventually to understand not only how a behavior is determined in part from genetic controls, but how these controls interact with experience (species-shared or unique) to produce functional actions when needed.

Summary

The opposed concepts of instinct and learning, or nature and nurture, as developmental causes for behaviors is a simplistic two sided division of a complex interaction of inheritance mechanisms, and species-specific and individual growth and experience. The concepts of hard and soft wired programming have some merit in this context as they reflect patterns in the needs of animals for some behaviors to be expressed effectively on first or only performance, whereas for various reasons other behaviors can be performed ineffectively on the first performance or early repetitions. Parental care in particular allows improvement through maturation and experience, with the parent maintaining the survival of the maturing, learning young. The association of environmentally important with unconditioned stimulii and the reinforcement of responses to environmentally important stimulii are effective behavioral improvers. In general hard wired behaviors, i.e. those performed effectively on the first or only performance, are needed in small, short-lived specialist animals with complex life cycles. Soft wired behaviors are functional for large, long-lived generalist animals with direct life cycles. The inheritance of hard and soft wired causation systems is not through simple Mendelian patterns, but is from the complex interaction of genes determining structures and the manner in which they are used or deployed.

CHAPTER 6. PETS, ZOOS, CIRCUSES AND FARMS: PERSONAL IMPACTS ON ANIMAL BEHAVIOR.

People affect the behavior of animals in many different ways. We should understand these effects, whether they are in our best interests and that of the animals, and if not, how they might be changed. The first step is to reduce these effects to categories reflecting different levels or types of effect, with potentially different types of action on our part to make what seem to be desirable changes.

A convenient division of human impact on animal behavior is into personal and impersonal levels. At the personal level we can consider such activities as having pets and operating zoos, circuses, and farms. In all these cases interaction between people and animals is at an individual level. The good zoos, circuses and farms have keepers, trainers and stockmen who are superb observers and naturalists, who love their animals and are committed to their care. We also impact animal behavior, or animal populations through their behavior, by environmental management or mismanagement. Here the impact is at an impersonal level, and is considered separately (Chapter 10).

The processes by which we manipulate the behavior of animals so that we can maintain homes with pets, or zoos, circuses and farms, can be divided as training, caging, domestication and taming. The divisions spread over a sequence of personal impact on animal behavior ranging from lifetime effects on one or a few individual animals, to animals in populations, and over generations.

Training

Training is the process by which we can control the behavior of an animal. It is, of course, a form of learning, and in recent years has advanced in respectability from the tricks of circus animals to the educational exhibitions of training procedures in some dolphinaria. Nowadays the career of animal trainer often requires professional qualifications as a behaviorist or animal psychologist, even though the trade secrets of trainers of hunting dogs and circus animals are still passed verbally from master to apprentice. There is an enormous range in the practical skills and theoretical understanding of trainers, whether they be amateurs with their household pets or professionals entrusted with the care of animals representing a large financial investment. The skills and theory needed are nevertheless the same. Behavior must be directed and shaped largely by operant conditioning. The theoretical underpinnings were shown by Breland and Breland (1966), but up-to-date details in concise form can often be found in manuals at dolphinaria (e.g. Schrage, no date). It is very noticeable that such large aquaria may provide educational material and reveal their methods, whereas traditional training agencies tend not to do so. Dog training schools provide practical instruction. They are the easiest entry into training for someone with no prior experience, but such schools may lack a convincing explanation of their success (or lack of it).

The essence of training methods is a consistent well trained trainer, immediate reward, and where necessary a long-distance secondary signal that a reward is due. The human trainer must be consistent so as not to confuse the animal trainee. Skinner boxes reducing such inconsistencies can be built for some animals and some of their behaviors (Figure 5.3). Reward must be provided within the memory span of the trainee, so that it can be associated with behavior performed. This is essential where a complex behavior is being shaped from simpler components, such as when a sea lion is taught to retrieve a ball in the water, climb onto a platform and toss the ball through a basket. A long-distance secondary signal such as a whistle is needed when the training covers areas distant from the rewarding person or automated machinery. This gets the animal's attention and is a signal of the reward earned. Whistles penetrate long distances in both air and water, and do not require the animal to keep looking at the trainer for a reward signal. Some animals respond to the secondary signal by immediate recovery of their reward whereas others, such as sheep dogs, can accumulate credits such as the intangibles of master-assistant satisfactions. The business of rewards and shaping must be adapted to the sensory and coordinative capacities of the species concerned, and also to variation between individuals of the species. Some individuals are not as clever as others and some can be contrary, moody or teasing. Also, with heavy animals the trainer needs to plan many aspects of behavior shaping, including his positioning and type of signal. Bad positioning can lead the animal to damage itself in falls, and signals must flow from one component to another.

Some trainers use punishment as a negative reinforcer to prevent some actions or induce aversive responses. It is not difficult to detect when punishment has been used excessively as it can generalize to produce overall submissive bearing and behavior of the animal. There may be good reason for using punishment to protect oneself if dealing with dangerous or large animals or assertive pets, but behavioral side effects can occur and need monitoring. There is also the ethical question of cruelty, and what is reasonable and unreasonable punishment. Refusing to give the reward for a partially performed action under training may be considered punishment, and a good trainer must be careful to refuse to give the reward when it is not merited. In contrast, scientists who set up experimental designs to electroshock drunken mice or starve vivisected sea slugs raise ethical questions, and should answer them.

The fairly widespread attitude that training is simply the production of entertaining tricks needs critical examination on behavioral grounds. The attitude is often extended to mean that animals should not be caged and trained for such entertainment. In contrast, zoo and dolphinarium operators have found that training can occupy the time of those species which are compulsively active under caged conditions, but whose cage habitat does not provide sufficient space or complexity for their activities. In species such as primates, canids, ursids and some ungulates, compulsive activity may reduce to pathological stereotyped pacing and rubbing, or hypersexuality. Morris (1966), Terrace (1979), and others report that such training and opportunities for self-expression of trained behaviors may be awaited and seized upon by the animals, in energetic ways which give us the appearance that they enjoy the experiences. It appears that where society allows the caging of animals, training can mitigate the impact, and even turn stress into healthful shaped activities. This is behavioral engineering (Markowitz and Stevens 1978).

It should also be noted that the young of some wild animals are provided with opportunities to learn by associating with parents and kin when they are engaging in certain activities, or simulating them. This is very close to training as we practice it.

Caging

When animals are brought into captivity, we can make very large changes in their habitat, the most obvious of which is that we constrain the space in which they must carry out their daily activities, i.e. we cage them. But there are other effects which may be quite independent of the available space or mixed with it. We constrain their choice of food and habitat material and the way in which they procure it; we constrain their choice of mates and selective pressures determining fitness; we inadvertently can culture their pathogens and create disease outbreaks; and we force the company of an alien species (ourselves) upon them (Hediger 1950).

The effects are most severe on those animals which in the wild actively and regularly range over long distances. They may be large and/or fast animals, which compounds the space problem, but the common feature appears to be an internal motivation for regular prolonged activity. This may be daily (home range patrolling) or seasonal (migration). Within a cage such ranging is very constrained. The activities of these species then degrade into prolonged pathological stereotyped pacing. Inadvertent rubbing and skin lesions follow. Some mammals, such as canids and ursids, can be kept caged in spite of the problem, especially if they are provided with activity regimes and training, but many far ranging birds can only be kept by wing clipping restraints. Long-distance fish migrants such as tuna and salmon often cannot be kept in aquaria, and dolphins must have training regimes. Even dolphins break into persistent picking at surface features of their tanks and they are liable to damage themselves. Smaller animals such as many tropical fish and birds have a long history of being raised in cages or aquaria, and skillful bird and fish breeders understand and provide good habitat as well as the other biological needs of their caged species.

Less mobile or even sedentary species also have spatial problems. These arise from the biological phenomenon of personal space with its components of alert, flight and defended distances, or its converse, the absence of some components of personal space in contact animals. The three components are staged around an animal as shown in Figure 6.1. The outer limit is the alert distance at which an animal responds to ecosystem changes which may represent a survival threat, usually a predator. Closer in is the distance at which the animal will break into escape behavior, and closer yet the distance at which it will turn and defend itself. When the walls of a cage are put within these distances for any particular species, then the animal may be in a constant state of alert or on the verge of breaking into panic escapes which cannot take it away from the people causing them. The problems are most serious for wild animals immediately after they are caught and caged, since fortunately many species can habituate. If so, their alert and flight distances may shrink for humans, and they start the process of becoming tame. Animals with a defended distance can be very dangerous to a zookeeper who must enter their cages. In contrast, some species are contact animals who may substitute their keeper for the contact stimulii that they seek or tolerate from conspecifics.

Notice in the diagrams of Figure 6.1 that the animal has been placed in that part of the cage farthest away from the

1. Personal Space Effect

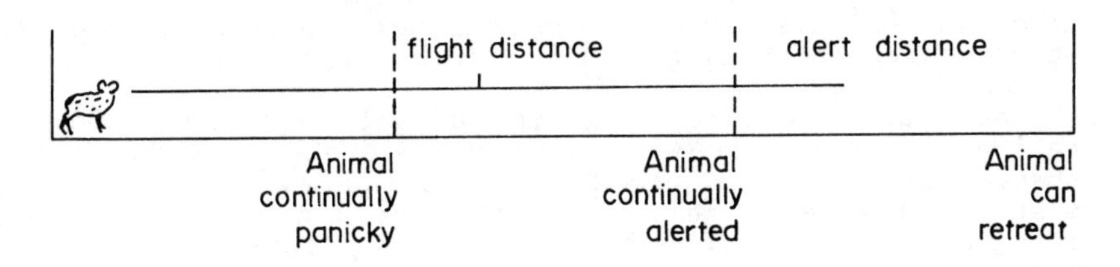

2. Home Range Effect

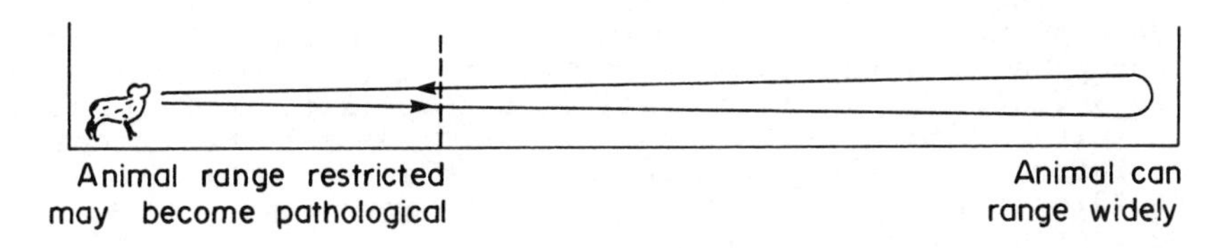

3. Territory Effect

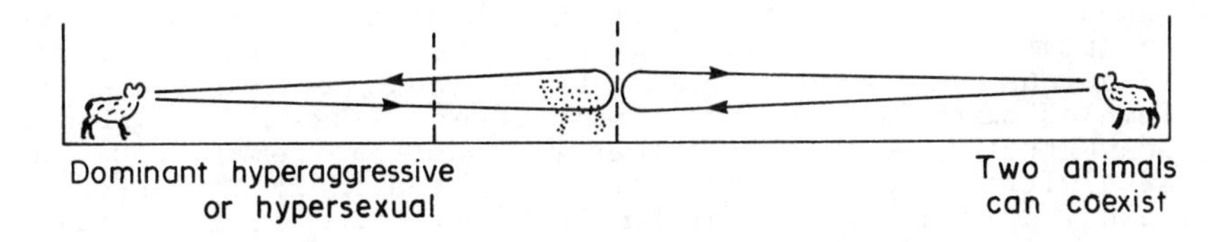

Figure 6.1. Caging effects on behavior. Too small or too simple cages (no hiding places) can produce undesirable behavior. Dashed lines indicate cage walls badly placed in terms of species' personal space, home range and territory.

source of disturbance. Even though it may habituate to alert and flight stimulii, it may still not be able to live well because it is constantly forced into retreat to the extreme end of its cage, or into a hiding place from which it can no longer perceive the disturbing stimulii. An animal's cage must be designed in conjunction with its potential to habituate to disturbing stimulii so that it can feed, mate, and engage in any other required behaviors without the constant production of panic escapes. A state of occasional alert might even be beneficial, and render the animal more interesting to the informed zoo patron.

The cage must also allow the animal to gather food and habitat materials in ways appropriate to its needs. Thus specialized feeders such as ant eating mammals and birds must be provided with live ants. This is an obvious case for understanding of material needs by zookeepers. Less obvious cases are continually being found as observers solve the problems of particular species. Thus ocelots present problems of self-mutilation as they strip their fur. Natural selection has programmed them to pluck their prey, and in the absence of pluckable food they turn on themselves (Barash 1979). A cage needs to be structured in ways appropriate to the locomotory, hiding and comfort needs of the animal. It may need sufficient complexity to facilitate activity, hiding or places for subordinates to escape a dominant, for sleeping spots, or the provision of nest building or toilet areas. There are many facets to designing a successful cage.

Successful caging ensures breeding. Unless the species can be bred, the keeper has not fully succeeded in his or her efforts. This presents a number of problems. First, there are limits on the number of individuals available for breeding, and their limited gene pool may fix unusual (and undesirable) characters, such as albinism, in the zoo stock. In mammals particularly, there may be complex social relationships between individuals, and certain males and females may not mate with each other. Goodall (1971) has shown that in wild chimpanzees at least one female would not mate with her younger brother, and screamingly protested the sexual interest of her older brothers. The incompatibility of Russian and U.K. giant pandas shipped to marital cages in both countries at great expense has been a topic of media humor for a long time (Morris 1979). The general problem may be either no opportunity to learn how to mate by watching others, or a programmed incest inhibition. There is yet another problem concerning the mating of socially organized mammals. In harem species such as baboons, males compete and fight for females as a resource, and in small zoo populations one male may monopolize all the females but be forced into constant fighting to maintain his rights, especially under crowded conditions.

Crowding in cages can have extreme effects as shown by the experimental rodent populations of Calhoun (1962), but normally these situations do not arise in exhibition type agencies such as zoos. Apart from inducing unhealthy conditions, crowding may distort dominant-subordinate relationships to such levels that the result is beautiful apathetic morons or hysterical murderous rat gangs. In contrast, one crowded group of mice made such an unusual step in learning that they modified and improved tunnelling behavior by inventing a rollable waste ball which could be pushed away, rather than each component being carried away one part at a time. Unhealthy crowding should no longer be expected in zoos where animals are kept for public exhibition, but is a risk in caging animals for scientific laboratories, especially where behavior is of little concern. In many countries nowadays the rights of animals are protected by Societies for the Prevention of Cruelty to Animals, and equivalent groups of scientists who themselves police animal quarters. The inspecting officers of such agencies and non-behaviorist users of laboratory animal colonies should be aware of the behavioral, as well as the physiological and visible, symptoms of illness.

In recent years there has been growing concern regarding the ethics of maintaining animals in cages, particularly the far ranging marine mammals, but also a number of other species such as the large eagles and other endangered species. There is also concern that the hunting activities of professional zoo suppliers may decimate stocks, since whole social groups or families may be killed to provide just one individual for a zoo competing for novel species. Many newly caught animals die en route to their eventual destination, or species which are caged following capture as wild young or adults may have a seriously shortened life span in captivity.

The counter argument is that making animals accessible to the public in an educational way provides benefits to people, and to the animal species concerned. A biologically informed public is open to espousing the ideals and goals of conservation. In addition, some zoos have become centers for the maintenance of endangered species and research into the means of managing wild stocks effectively.

Opinions on the topic are diverse, and involve balancing the various factors. Not the least of these is the capability of a particular agency to maintain animals in cages, and its motivation for doing so. The monkey supply house which provides specimens for vivisection, and the exporting shipper in an impoverished tropical country are very different social institutions from a zoo or dolphinarium with an educational program. They are interlocked, however, in a network of social arrangements. If you accept one, you may be acquiescing in the others.

It is usually easier on both the animal and people if the animal being kept in a cage has itself been reared under caged conditions. The advantage is that it becomes used to the habitat and its regularly visiting people, and may even imprint on such regular visitors, thereby treating them as conspecifics. A captured wild infant may also be imprintable, although the degree to which such learning while an infant influences adult behavior is variable. For large fierce animals substantial risks remain for a trainer or keeper since their size alone makes them physically dangerous. However, the level to which friendly relationships can be built between people and large animals is shown in modern dolphinaria and in many zoos and circuses. They can occur; but that does not mean that everyone can attain the same relationship with a particular animal. In dolphinaria there are many people who cannot induce the whales to perform well. Herbert Terrace (1979) found that the chimpanzee Nim rejected some proposed research assistants for sign language teaching.

Other aspects of caging, such as the prevention of malnutrition from the food supplied or the treatment of diseases and injuries, are more the subjects of physiology and veterinary medicine than those of behavior.

In conclusion, with regard to zoos, aquaria and other animal keeping agencies such as stables and pounds, it should be noted that adequate care can be provided by understanding the behavior and other aspects of the biology of the species concerned. In addition, enlightened zoos and aquaria provide facilities for research and developing better care of the animals, as well as fundamental research.

Domestication

Domestication is the most extreme form of our personal impact on animals since it involves bringing not only the care and feeding of the animal under human control but also breeding. Under domestication, the mating of animals is controlled so that a stock is produced with qualities desired by the human breeders.

It is particularly important for the behaviorist to understand the nature of domestication, since domestic animals have had profound impacts on human cultures. Domestic animals have contributed to human affluence, one of the determinants of human behavior in different cultures. We ourselves, in our recent evolution, may have imposed domestication-like cultural controls on our own breeding (Morris, 1969). Understanding the behavior of domesticated animals may be useful, not just in better use of our farm resources, but in understanding ourselves.

The range of qualities produced by controlled breeding is enormous, since at one time or another people of diverse cultures have bred animals for a number of general uses, but uses also specifically adapted to the culture of the people concerned. Table 6.1 summarizes the nature of uses for which animals have been bred. Many of these uses will have been secondarily derived in the sense that the original domestication of a species then makes that form available for many uses, whatever its initial role was in the society which started the domestication process. These current uses include food, protection, materials production and handling, transport, comfort and inclusive fitness. These categories follow those adopted for ethogram construction (Chapter 3). Most of these categories are self-explanatory, and the variety of animals serving the various functions well-known. Two categories need special mention. Comfort as a use for animals encompasses all uses of pets, the role of which is not material but the provision of company, contact and affection. People and some animals which make good pets are capable, between species, of expressing the group organizational behaviors which normally maintain kin groups together. Although mixed species flocks of birds and schools of fish are occasionally seen, the species cncerned are often closely similar or the relationship is functionally symbiotic. The bond between pets and people spans substantially different species, and is individual. The effect is to produce a group of two or more bonded individuals who regularly seek each other's company for the intangible benefits of association and comfort. They are friends in much the same way as were the chimpanzees Leakey and Mr. Worzel (Goodall 1971).

The category "inclusive fitness" encompasses animals whose use increases the fitness of humans individually or the groups in which they live (nations, races, etc.). Thus animals are bred for use in warfare and colonial exploitation, and individual domination of one person over another. Even sporting events where victors receive rewards and recognition can be brought into this category.

The uses in general are to meet human needs for materials and energy in the form of food, but also to provide energy in the form of work. Thus the uses supplement the basic requirements of any species in the food web of an ecosystem, but have been expanded to provide mechanical energy and the non-energetic intangibles of comfort.

In spite of the very great number of uses to which domesticated species have been put, there are surprisingly few species which have responded to the artificial selective pressures imposed by humans. Table 6.2 lists the common species. It is very noticeable that most are mammals or birds, and that the ungulate order Artiodactyla has produced a large number of species with various uses. The bird order

Table 6.1. Uses of domesticated animals.

Use type and subdivision	**Animals**
Food	
Slaughtered	Cattle, pigs, sheep, oysters, trout
Processed	Cattle (milk), hens (eggs), bees (honey)
Protection	Dogs, geese
Materials	
Production	Silkworms (silk)
Construction	Elephants
Transport	
Passengers	Horses
Freight	Oxen, pigeons
Comfort	Dogs, cats
"*Inclusive fitness*"	
War	Elephants
Domination	Dogs, horses
Sport	Horses, falcons, pigeons

Galliformes has also produced many domesticated forms. There is a pattern here suggesting that certain types of animals have qualities which allow them to respond to the selective pressures of controlled breeding.

It could be that other animals can be domesticated but have not been in contact with human cultures when need and

Table 6.2. Taxonomy of common domesticated species.

Mammals

Order Perissodactyla
  Family Equidae - horse, donkey

Order Artiodactyla
    Suidae - pig
    Camelidae - camel, llama
    Cervidae - reindeer
    Bovidae - cattle, sheep, goat, yak

Order Carnivora
    Canidae - dog
    Felidae - cat

Order Rodentia
    Muridae - rat, mice
    Cavidae - guinea pig

Order Lagomorpha
    Leporidae - rabbit

Birds

Order Anseriformes
    Anatidae - duck, goose

Order Galliformes
    Phasianidae - chicken, pheasant, quail, peafowl
    Numidiidae - guinea fowl
    Meleagrididae - turkey

Fish

Family Salmonidae - trout
A great variety of small tropical fish as exhibition breeds
  Siamese fighting fish

Insects

Order Hymenoptera - honeybee

Order Lepidoptera - silkworm

Order Diptera - fruitfly (breeding for scientific research)

Order Coleoptera - grain beetle (breeding for scientific research)

Table 6.2. continued.

Molluscs

Bivalves - oyster, mussel

Bacteria

A variety of forms being bio-engineered for metabolic purposes, e.g. insulin production

opportunity arose. Also, genetic engineering may soon change domestication and breeding potential. Figure 6.2 maps the distribution of the earliest fossils or records of animal use suggesting domestication, i.e. breeding control. It is apparent that some areas of evolving human society favored the development of domestication, especially in the Middle East. There were lesser areas in South America, however, and areas where single breakthroughs may have occurred, as with dogs in North America.

The question now is: what are the processes by which certain species are domesticated? Much of such a description involves human biology and behavior in prehistoric society and is beyond the scope of this book, but informed fiction provides credible accounts (Lorenz 1955). There is one component to the domestication process which the behaviorist can contribute. It was summarized by Hale (1969), who developed the concept that behavioral characteristics of certain animals favor their responses to selective breeding. Table 6.3 is a modified version of Hale's concept.

After considering Hale's domestication-favoring behaviors the general conclusion is that many behaviors interact to facilitate the process. Animals living in the wild in mixed sex flexible groups with a ranking structure, and having social signals mediated by behavior, can easily domesticate. So can those with an imprinting period by which they learn their own parent and hence species image. They present the opportunity for humans to take over the role of dominant parent and thus facilitate individual control. This in turn allows the control of breeding, so that the gene pool can be manipulated. In addition to the other characters of the species, domestication is favored if the young develop quickly to relative mobility and feeding (are precocial), have a juvenile learning period, have a short flight distance to man (which means the ability to escape from him effectively in the wild unless very close), and are

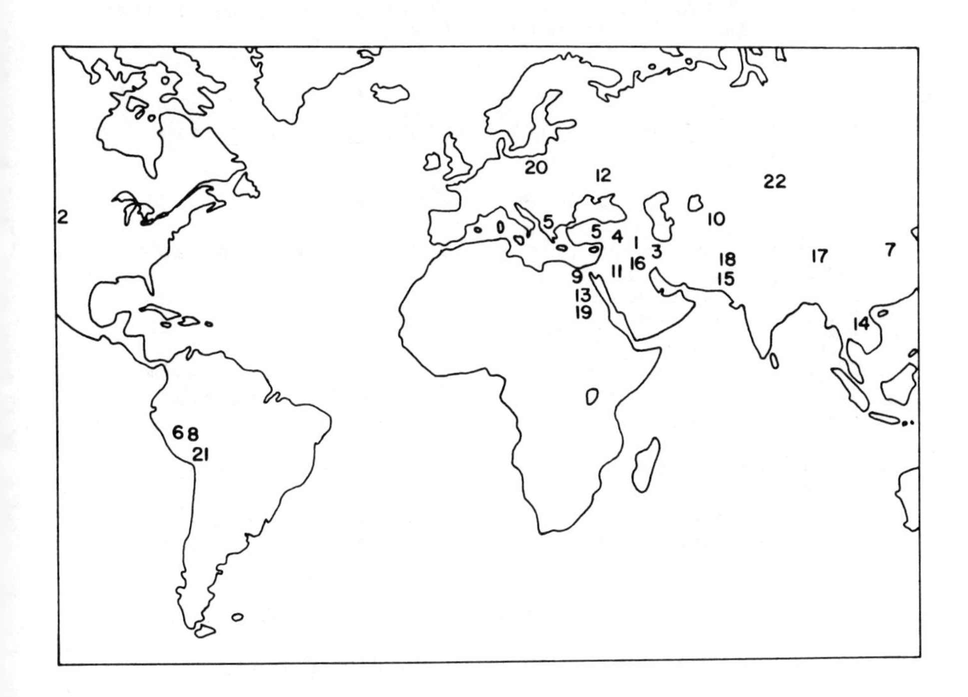

1. Sheep 8500 B.C.
2. Dog 8400 B.C.
3. Goat 7500 B.C.
4. Pig 7000 B.C.
5. Cattle 6500 B.C.
6. Guinea pig 6000 B.C.
7. Silk moth 3500 B.C.
8. Llama 3500 B.C.
9. Ass 3000 B.C.
10. Bactrian camel 3000 B.C.
11. Dromedary 3000 B.C.
12. Horse 3000 B.C.
13. Honey bee 3000 B.C.
14. Banteng 3000 B.C.
15. Water buffalo 2500 B.C.
16. Duck 2500 B.C.
17. Yak 2500 B.C.
18. Domestic fowl 2000 B.C.
19. Cat 1600 B.C.
20. Goose 1500 B.C.
21. Alpaca 1500 B.C.
22. Reindeer 1000 B.C.

Figure 6.2. Distribution of earliest records of domestication.

otherwise rather tolerant of environmental intrusions, i.e. are adaptable to environmental changes, both long-term and short-term, and are omnivorous or opportunistic feeders. Hale's character of limited agility favors domestication of animals which do not need to be constrained, but is in conflict with the requirement for short flight distance to

Table 6.3. Behavioral characteristics favoring domestication (modified from Hale 1969).

Favorable characteristics	Unfavorable characteristics
I. Group structure	
A. Large social groups (flock, herd, pack, etc.) and frequent individual exchanges between groups	A. Small almost permanent kin groups
B. Ranking organizational structure, with leader (or initiator)	B. Territorial or caste (colonial) organizational structure
C. Mixed sex groups	C. Sex groups separated
D. Social signals by movements and posture	D. Social signals by badges (color or morphology)
II. Sexual behavior	
A. No sexual bonding	A. Pair-bond matings
B. Short flexible courtship	B. Long stereotyped courtship
III. Parent-young interactions	
A. Imprinting of young to parent and learning of species image	A. Inherited recognition of species image
B. Single parent raises young	B. Two parents raise young
C. Raising parent learns young (can foster)	C. Raising parent recognizes young by inherited characters
D. Precocial young	D. Altricial young
E. Moderate learning juvenile stage	E. Very long or very short learning period

Table 6.3. continued.

Favorable characteristics	Unfavorable characteristics
IV. Responses to man	
A. Short flight distance	A. Long flight distance
B. Little disturbed by man or sudden changes in environment	B. Easily disturbed by man or sudden changes in environment
V. Resource adaptability	
A. Opportunistic	A. Specialized
B. Omnivorous	B. Specialized food requirements
C. Broad habitat tolerance	C. Narrow habitat tolerance
D. Limited agility	D. Extreme agility
E. Direct life cycle	E. Complex life cycle (larval stages)
F. Moderate size and power	F. Very large (powerful) or very small

man, which implies considerable agility. Many factors must be balanced and some conflict with each other.

In practice, only a few species have shown a satisfactory balance of these factors. Historical attempts to domesticate musk-ox and species for aquaculture show the difficulties in working with species for which the balance is not particularly favorable. Musk-ox are just too specialized as cold climate grazers. Among aquaculture species only a few such as rainbow trout have been domesticated to the level at which particular strains are being produced for rapid growth, pan size, and attractiveness when presented on the plate. Some tropical fish are bred for aesthetics, rarity and behaviors such as fighting.

Hale (1969) also outlined an analogy between the sequences of change brought about by natural and artificial selection (Table 6.4). This reflects the effects of initial

Table 6.4. Comparison of natural and artificial selection processes, demonstrating the sequence leading towards present day levels of domestication (modified from Hale 1969).

Natural Selection	Artificial Selection
A. Environmental changes generate new ecosystems for novel adaptations	A. Human cultural change allows close symbiotic association between animals and humans
B. Adaptive radiation by many species to many different trophic niches	B. Taming and controlled breeding of many species with overlap of uses
C. Selection between species	C. Selection between species for particular uses
D. Improved adaptiveness under competitive pressure	D. Improved specialized utility (strain breeding)
E. Highly specialized relict forms or species becoming maladapted and extinct	E. Limits of responsiveness reached, or maladaptive forms become frequent (strain discontinued)

domestication as a response to the early creation of a human derived ecological niche in agricultural or hunting societies, following which the radiation of strains by consciously or unconsciously controlled breeding allowed stocks to survive in many different societies. Nowadays breeding control is absolute with stud farms and artificial insemination, and has led to the genetic irresponsibility of some dog and cattle breeders producing deformed and sickly breeds.

The products of domestication in widely divergent species show convergence (Tinbergen 1951). Thus animals bred for food tend to become bulky with muscle and have short legs. Those bred for speed are long legged and lean. Those bred for freight transport are large and bulky (again from muscle) and have considerable power. Those bred as pets retain infantile features that function as sign stimulii releasing care and affectionate behavior from humans.

Taming

There is one change in animal behavior which affects the controls and constraints of training, caging and domestication. This is the human aided change of taming by which the aversive or aggressive behavior of animals is reduced. A tame animal can be trained more expeditiously in that the trainer has freer access to the animal. A tame animal can be caged more efficiently in that it can be properly cared for without risk to itself or keepers. A domesticated animal can be farmed more economically if it is sufficiently tame that it can manage itself without expensive constraints or artificial habitat.

Figure 6.3 illustrates the relationships between taming and caging, and training and domestication. The initial process by human societies in bringing animals of wild stock under their control must have been either by the physical processes of caging (or hobbling) or the psychological processes of imprinting on captured infants. The constrained animal habituates to humans not preying upon it, or it imprints to a human foster parent. Both must have occurred as hunting and gathering fathers, mothers and older sibs captured newly born or hatched animals and handed them over to younger sibs as playthings. The tamed young animal was then available for training and breeding, so that the processes of use and domestication were initiated.

At this point the limited gene pool phenomenon, known as the "founder" effect (MacArthur and Wilson 1967), will influence the stock and its capability for further shaping to human needs. When only a small stock is available for breeding, the descendants of the stock are limited in their adaptiveness to the contents of the original animals' gene pool, which will be missing some components of the entire species pool, or may have unusual alleles which can be fixed readily by inbreeding. The problems can be reduced by breeding with wild stocks (as Eskimos stake out their husky bitches to be bred by wolves), introducing newly captured wild individuals, or crossbreeding specialized strains.

In species for which the human use is for exhibition purposes as in zoos, the founder effect can be serious. Few individuals available for breeding means a limited gene pool, and pressure to collect more specimens from the wild. The founder effect appears to have been less serious for useful forms, as large sections of the stock appear to have been taken over. The missing or gradually shrinking wild stock of such forms as horses, pigs and cattle may be due, of course, to their inability to adapt to human derived selective pressures on the ecosystem to which they were adapted.

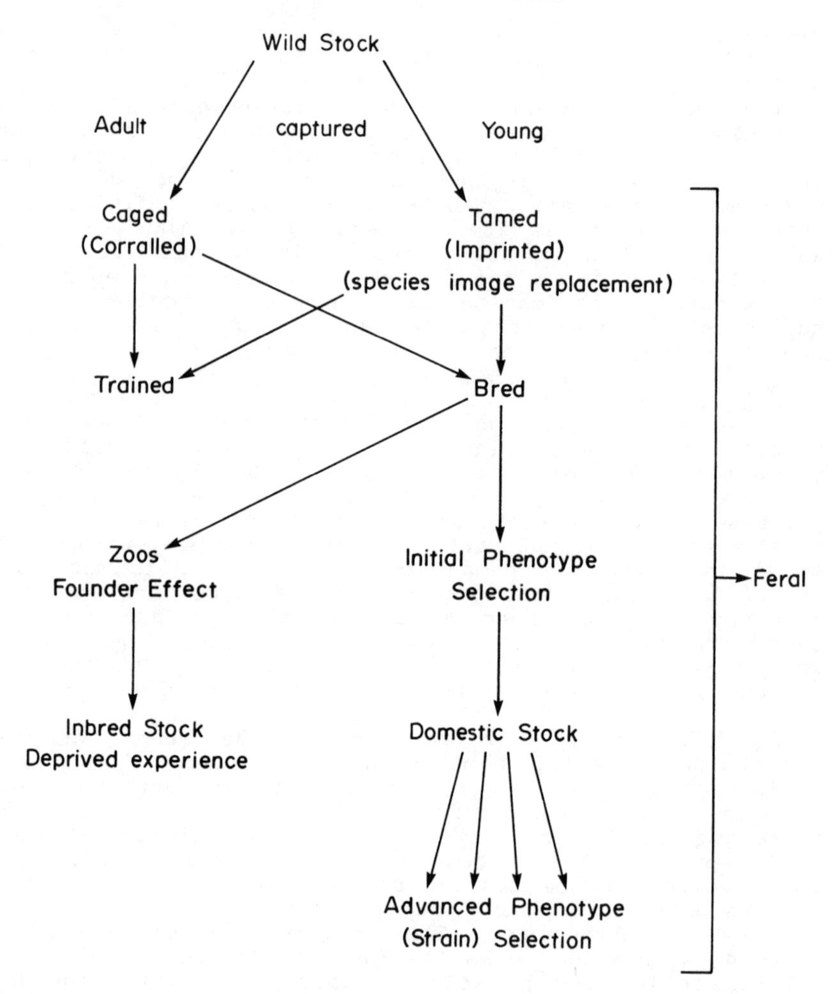

Figure 6.3. Relationships between taming, training, caging and domestication.

There is another use of the concept "tame". Many free ranging animals associate with us in our urban and agricultural societies. They are symbiotic with us in that they have adapted to our ecosystem modifications, which are extreme in the downtown areas of cities and around industrial plants. Some of these species are tame only in the sense that their agility allows them to escape from our relatively close approach. In other cases, such as that of mallard ducks living in urban parks, their learning ability for human body language allows them to distinguish between people approaching with harmless or harmful intent.

Farming and aquaculture

Present day farming and aquaculture attempt to maximize the output of whatever products the various domesticated, or semi-domesticated, species provide us. But they continually run into problems of animal behavior, particularly where the animals are constrained -- as in hen batteries and fish hatchery tanks. The effects on the animals are similar to, but far more intensive than, those appearing under crowded zoo conditions. However, where domestication has produced placid animals the effects of constraints are not as severe as on wild animals. Nevertheless battery hens do rub patches bare, and hatchery trout scrape the scales from their sides, as they insist on moving around in their confined space.

More subtle behavioral effects are well-known to farmers, and they may often crop up unexpectedly with changes in farm mechanization. Thus the follow-the-leader behavior of a herd of cows to their twice daily milking is a ranking behavior that farmers must accommodate, if they wish milking to be peaceful, productive and a minimum-labor activity. It is surprising how cooperative animals can be if their behavior is taken into account. Cows led to a new milking parlor after years of traditional cowshed practices have been known to order themselves effectively as quickly as the second day of using the parlor.

Similarly, with species which are more difficult to domesticate, some of the difficulties are behavioral. The problems in domestication, and improving the stock of potentially useful species, have become very noticeable in recent years with the extension of aquaculture to marine invertebrates. Many species of tropical fish have been domesticated for food and recreation over the centuries, but invertebrates present new problems. Many domesticated fish are territorial forms with little need for extensive ranges, and so are easy to raise under caged conditions. However, domesticating such species as trout, salmon, lobsters, oysters and other shellfish is a very different procedure.

Trout and salmon are far-ranging, migratory species, and salmon particularly need considerable space for ranging, and their diving escape response from surface predators. There has been considerable success in hatchery rearing the less migratory trouts, to an extent where particular hatcheries using controlled breeding appear to be producing desirable stocks which are sought by other hatcheries. A typical domesticated animal is being produced -- fat, and placid (but to some people bland in taste). In aquaculture terms, a desirable stock of trout must grow to pan size as quickly as possible, with minimum damage to itself, minimum disease and minimum disturbance from stimulii generating dart escape responses using up energy.

The problems with shellfish aquaculture can be very different. There is a long way to go, for example, to convert the skinny wild mussels of the Pacific west coast into the farmed plump morsels favored in Europe. Shellfish aquaculture is complex since many species have an indirect life cycle with larval stages. Where these larval stages are planktonic, as with most mollusc and crustacean forms, their needs for particular habitat, and the behaviors of the larvae, must be accommodated under domestication. The planktonic drift stage is in itself not that much of a challenge. There is a long tradition in marine biological laboratories of culturing larvae, at least in small numbers, for scientific purposes. Such procedures can be scaled up to raise larvae by the millions. However, the problems come, in shellfish aquaculture and controlled breeding, with the transfer period as the drifting larvae settle from the water column to the habitat required by the adult. As the larvae settle, they must change anatomy, physiology and behavior, remaining as they do so a viable individual capable at least of surviving predators. This is difficult. Many such larvae have a pre-set time during which they must make the changeover. They also have a sensitivity to the right conditions, so that they test by their antennae and other sense organs, and can reject or accept whichever is appropriate. If they don't find the right habitat in time, they disappear.

Until this settling stage is controllable, the aquaculture and domestication of larval-producing shellfish species cannot make much progress. There are other problems including the similarity between different species (such as the scallops) at their young stages, so that mixed stocks may be unknowingly cultured, with an unknown proportion being less useful than the rest. In spite of these problems there is considerable development now in aquaculture. Knowledge of behavior of the various species will do much to overcome the practical difficulties.

Summary

Humans impact on the behavior of animals at personal and impersonal levels. At the personal level the types of impact can be categorized as training, caging, domestication and taming. Training brings the behavior of individual animals under human control and can be used beneficially in the work use of such animals as sheep dogs, and the caging for human purposes of compulsively energetic animals with considerable learning ability, e.g. primates and dolphins. Caging refers to practices which constrain the movements of animals in zoos and dolphinaria, but also to the constraining of fish in aquaria and game in compounds. Successful caging requires knowledge of the ecosystem adaptations of the species concerned, and the animals' ability to accept constraints arising from the cages devised for them. Domestication is the extreme form of human impact as the breeding of the individual is brought under control and the stock is modified for efficiency in whatever use is found for the animals. Domestication may only be practical for a limited range of species which have particular group living behaviors favoring placidity and constraints, and may be subject to founder effect gene pool limitation. This last factor seriously affects individuals in restricted populations of zoo animals. Finally, taming is a phenomenon by which aversive or aggressive responses to people are reduced, and may either be learnt individually or appear through inheritable modifications under natural selection in human impacted environments or under artificial selection for domestication.

CHAPTER 7. PHYSIOLOGICAL (PROXIMAL) CAUSES.

As mentioned in Chapter 1, when we try to "understand" behavior we have to consider two aspects of "understanding". We can understand how an animal does what it does, and we can understand why it does it. Answers to "how" questions are answers in terms of proximal causation. What are the immediate causes of the behavior in terms of motor organ action and physiology? For example, what is occurring in each compartment of the behavior system (Figure 1.5) as the animal senses something, processes the information, and then responds with muscular action? There are more lengthy processes involved as well. The animal may have only been primed for that particular behavior as a result of seasonal or one-shot hormonal flooding of sexual maturation. The biochemical processes of ontogeny creating structures and their functioning are part of the answer to "how" questions. The material in Chapter 5 on programming by hard or soft wiring also provides some of the answers. There are many levels of understanding proximal causes.

"Why" questions lead to answers phrased in terms of the survival value or adaptiveness of the behaviors performed. Why should a *fischeri* lovebird carry nest material tucked in its rump and tail feathers instead of employing another equally improbable way, or the more comprehensible way of using its beak? Presumably it does so because it can carry more pieces at a time and thus reduce its energy expenditure in the inevitable benefit-cost ratio that natural selection imposes on all animals all the time.

Answers to the "why" questions are answers in terms of ultimate causes, since natural selection is considered the ultimate shaper of animal behavior. An introduction to ultimate causation is presented in Chapter 10. This chapter now considers proximal causation in ways intended to interface with the various components of the topic. Neurophysiology, neuroendocrinology, brain function analysis and other components are all complex subjects. They attract their own specialists, and they each provide information relevant to understanding behavior.

The presentation to be followed was formatted in Chapter 1 with the model of the behaving animal (Figure 1.5). The model presented three black boxes: the sensory system, coordination (neuroendocrinal) system and motor system. Now we shall explore the contents of each black box to some extent, but each time we will encounter the inevitable situation of finding yet other black boxes nested inside, all awaiting individual decisions whether we will try to "understand" further.

Sensory systems and information reception

Energy forms and their animal sensors

Out there in the environment there are many forms of energy flowing through the ecosystem. The animal that can "bug" those energy flows, in the cold war sense of listening in, has sources of information which can help it maintain itself alive. It can sense the presence of food, predators, habitat, itches, time to be inactive, a mate, kin or non-kin.

Energy forms and the sensors for them are listed in Table 7.1. There are three divisions, each containing interrelated energy forms. However, we must exercise some caution. Humans receive information from all the energy forms listed, or the forms are sufficiently close to what we detect that we can understand how other animals with slight modifications of our common sensory systems can tune in to a range somewhat different from ours. What if there are energy forms beyond our present comprehension? In the past we did not understand electricity, gravity and pressure, and a congenitally totally blind person has no personal experience of light and vision. We must remain alert to similar possibilities in such fields as sensitivity to high voltage transmission lines and what is called Extra Sensory Perception (ESP), both contemporary subjects of investigation. But it should be noted that investigation of the latter is a field traditionally open to technical hoaxes, crackpots and charlatans.

In Table 7.1 only discrete sense organs are listed. For some of the energy forms, e.g. electricity and heat, there appears to be a diffuse sensitivity scattered through the body cells, or some of them. For other energy forms, the sensitivity may be diffused in the form of specialized individual cells or in cell clusters. These may be scattered through the body, or located in regions where the energy form is likely to be contacted. The chemical foot sensors of many insects are an example of the last. The presence of distinct organs shows that additional sensory advantages can be obtained by organ shape and mobility. Thus vertebrate ears and eyes are complex organs allowing

Table 7.1. Environmental energy forms and their sensor organs in animals.

Energy form	Sense organs
Electromagnetic and thermal energy	
Light	Eyes
Ultraviolet radiation	Eyes
Infrared radiation	Eyes
Thermal	Thermal pits
Electric	Ampullae of Lorenzini
Magnetic	Magnetic crystal sites
Mechanical energy and force	
Sound and sonar	Ears
Gravity	Ears
Inertia	Ears
Touch and vibration	Skin
Pressure	Skin
Chemical	
Taste	Tongue
Smell	Nose
Humidity	

sensitivity to a range of wave frequencies and other properties of the energy form, but they are also mobile receivers capable of deriving directional information about the energy source. Such organs may receive a variety of information from several properties of the energy form, and some sensory cells receive and respond to several energy forms, e.g. the human retina will generate visual sensations if the eyeball (when closed) is pressed.

Sense organ concepts

In order to explore the ways in which sense organs can adapt to receive ecosystem information, we need to consider various properties of sense organs.

There are a number of common properties of sense organs regardless of the energy form to which they are sensitized and whether the energy is derived from inanimate environmental or habitat sources, or animate sources of predator, prey or conspecific.

There is a minimum level of sensitivity, i.e. a threshold to response. There must be sufficient energy received to trigger the nerve response by which information is sent on its way (transduced) into the animal's transmitting and coordinating system. This sensitivity varies with the energy form and between species. There may be the ultimate of sensitivity to a few molecules only for some smells. Relatively large amounts of light may be needed by daytime animals compared to nocturnal visual hunters such as owls.

With wave energy forms such as electromagnetic and mechanical energy, only a limited range of wave frequencies may be sensed. Thus we sense electromagnetic wave frequencies from 380-760 nm, the wave band we call light, but other animals such as bees sense down into the lower frequency ultraviolet level of the spectrum. We sense mechanical wave frequencies between 20 and 20,000 Hz as sound (and this ability changes with age), whereas bats sense up in the 100,000 Hz range, and bees have three sensory windows at 0-50, 250-350 and 1500-3000 Hz with different organs for each range.

Within an energy form, there may be several components which can be sensed in different ways. Thus with light, wave frequencies may be distinguished as different colors. The normally random directional oscillation of the waves may be polarized more or less into a single plane at dawn or dusk, for instance, and this plane sensed. Light has direction, and the direction of origin can be detected by appropriate construction of shades for the sensor (e.g. eyes in sockets), and organ mobility.

The capacity of a sense organ can be determined in part by field observations of the uses to which it is put. For example, the eyes and ears of owls, as expected from their nocturnal habits, are exceptionally sensitive organs in terms of low thresholds, but also in localizing the direction of the stimulus. But such information gives only general guidance to capability. Experimental analyses are needed and these can draw upon operant conditioning techniques. Table 7.2 illustrates the sensitivities of various fish to a wide variety of environmental energy forms as derived by Bull (1957) using an L-shaped conditioning tank, one of the early systematically used Skinner boxes.

This list shows yet another property of sense organs, that of discrimination. Bull commonly expresses discrimination as a percentage change within the detectable range or from a starting level of perception. Some of the sensed information did not lend itself to quantifying

Table 7.2. Summary of sensory discrimination in fish, as determined by Bull (1957).

Sight

Color
Brightness
Form, cards with letters U.E.W.L.
Size
Number - groups of 2 and 4 letters
Position of objects

Sound

Range - 16-7,000 cps
Intervals - .25-1.33 octave at different ranges
Recognize sources of two sounds
Whistles associated with food
Human voice stating "Adam" or "Eve"

Smell and taste

Species discrimination by odor
Individual discrimination by odor
Plant rinses
Pollutants - phenol
Mussel extract - 0.00075%
Salt, sour, sweet and bitter substances
NaCl - 0.00023%
Acetic acid - 0.15%
Saccharose - 0.00072%
Quinine - 0.0025%
Natural sugars from saccharine

Current flows

Changes within order 5-40 cm/min, provided visual reference point

Directional changes 90° or more but not less, regardless of fishes orientation to the current

Ph

0.04-0.10 in range about 7.0

Table 7.2. continued.

Salinity (seawater)

Generally 0.5^{o}/oo
Gobius flavescens 0.06^{o}/oo

Temperature

0.03%

Pressure

0.5 cm decrease, 1.0 cm increase

discriminatory ability, such as the distinction between the words "Adam" and "Eve".

A key question in this context of discriminatory ability is whether the discrimination is actually used, and when. Lists of proven capability are no substitute for demonstration of when the capability is actually used.

Interfacing concepts (transduction and filtering)

Sense organs also have a transducer function. They must convert information received into a form which can be transmitted within the animal body. In so doing they must retain informational content and reduce noise from unnecessary information (and not introduce any new noise from their own transduction). They must act as filters for the maelstrom of energy perpetually battering the organism so that only information with survival value to the animal is actually transduced and made available for response.

This filtering effect is achieved in a number of ways. The first of these is the threshold property already mentioned; a minimum quantity of energy reception is needed to provoke the transduction to the nervous system. Second is the phenomenon in which specific information, such as a temperature level, is sufficiently well correlated with a complex environmental situation, such as the right time of year for spawning, that specific information can serve as a sign stimulus for the overall situation. Such sign stimulii then trigger a set of nervous-coordinative-motor events leading to behavior adaptive to the whole situation. This complex of sign stimulii and set motor response was originally conceptualized as an Innate Releasing Mechanism (IRM) producing an FAP (see Figure 1.1). Much has been added to the original concept, including such components as heterogenous summation of sign stimulii (Fabricius 1950). In these conditions, several sign stimulii may separately or by summation release the adaptive behavior. Fabricius quoted right temperature, vegetation and river gravel for pike spawning, but suboptimal values of one of these could be added to by increasing the stimulus of the others. The gist of the concept is that many environmental situations to which animals are adapted are complex, but can be represented by one or more stimulii which function as signs of the whole situation. Once the sign is received by the sensing animal at sufficient intensity, or several signs are summed in some way, a motor program is initiated which releases behavior adaptive to the originating environmental situation.

The filtering level may vary and should not be considered immutable. However, it may be difficult to determine if the changed filtering capacity is located in the sense organs or brain. The electro-recording of sense organ nerves is an experimental technique allowing analysis of what information passes the filter.

Unfortunately, some environmental situations are too complex to be reduced to single or even summed sign stimulii of this type, and the releasing stimulus appears to be a complex configuration of information the component parts of which must be formatted in a certain way. Thus recently hatched chicks may either freeze or ignore a hawk silhouette experimentally moved overhead, depending on which end moves forward. They will ignore the silhouette if the long tail end is forward and the model resembles a long-necked non-predatory flying goose. The configuration of shape is an important signal to the chicks, but it must be included in the even more complex configuration of movement relative to shape. Configurational stimulii are important in the visual mode of perception where landmark orientation is common, but there are analogues in auditory sensing if not in other modes such as olfaction. In auditory sensing, many vertebrates respond to the calls and songs of conspecifics. On analysis of sound spectrographs these calls have been shown experimentally to comprise complex essential configurations in some, and relatively simple signs in others (Figure 4.1).

Sensory adaptiveness

Every animal is loaded with a battery of sense organs, and there are several patterns to this loading. Animals are equipped to receive information which has survival value to them, and hence information about the behavior of a species will indicate the nature of its sensory capacity. Sensors are deployed where they are most effective, i.e. where the stimulii are most likely to be received. This is usually at the head end, the end which moves forward, since it is usually more important to an animal to sense what it is coming to rather than what it has already passed. There are exceptions, of course, in animals which are frequent prey species. These animals may have mobile front end sensors which can be directed to the sides or back, or even backwardly directed sensors such as the pressure sensitive anal circi of cockroaches. Radially symmetrical animals such as coelenterate plankton may have radially arranged sensors, and oddly shaped forms such as scallops may have eyes arrayed around the forwardly oriented shell gape.

Animals may be responsive to simple component signs or complex configurations, but we can expect that they will be responsive to any energy form which can carry information of survival value to them. We can expect animals to have subtle powers of discrimination to gain useful information. Particularly with regard to invertebrates, understanding sensory capabilities requires some imagination. It can help to imagine oneself in the animal's world of environmental energy fields, whether they be derived from inanimate sources, predators, prey or conspecifics.

Animals have evolved many different sense organs to receive similar information, e.g. the eyes of cephalopods and vertebrates. They have frequently converged on similar sensory adaptations. Also, with the apparent plasticity of sensory adaptation, redundancy is common. Animals can respond adaptively to environmental situations through the perceptions of several sense organs, e.g. salmon can spawn in turbid water where their visual sense is very limited and presumably where they are dependent on their lateral line system to detect the nearby courting other sex.

Motor systems

The production compartment of the modelled behavior system comprises the motor organs. Most of these are muscular, with a lever arrangement generating forces on the habitat so that the animal is propelled and produces its behavior. However, even the muscular organs can be divided into two categories, with the behavior of each substantially different and having different adaptive potential. The two categories of organs or motor systems are locomotors and contractors. Locomotor organs are those which propel animals through their habitat, whereas contractors are the organs which by shortening bring about relative movements of body parts without producing locomotion. Muscles may be combined with other cell and organ properties to build and deploy motor organs in ways which produce yet other forms of behavior. Thus many animals call, sing or otherwise vocalize as a result of muscular action forcing air through stringed organs, or rubbing combs together. Other animals use muscles to squirt species-specific or repellent chemicals, or otherwise apply glandular secretions to the habitat. In some cases, other cell or organ properties without muscular action produce behavior either in the deploying species or in the animals associated with it. Thus some species bioluminesce, generating light patterns, and others electroshock their prey. These are non-muscular behaviors.

The last few categories show a result of natural selection in an information loaded world. Some animals have become not just passive receivers of information, but are actively transmitting information themselves. Moving animals inevitably add to ecosystem information, and other animals which might be prey, predators, mates, competitors or cooperators benefit by receiving information about their movements. But the other categories of motor systems show an evolutionary development far more significant. Callers, smellers, lighters and shockers are actively transmitting information for one purpose - to influence the information received by other animals, and so manipulate their activities. Even moving has been brought into information transmission and behavior manipulation, as animals have evolved displays signalling threat, courtship or parent-is-here, or they have evolved stalking and cryptic behavior which minimizes normal information transmission.

In some cases, such as with potential mates, the receiving animal benefits by responding to the information received. In other cases it may not benefit, as in the interaction between predator and prey, where an evolutionary arms race balances adaptations. In cases of competition and threat, the loser is just that; he or she has lost since his or her behavior has been manipulated by information transmitted by the winner.

Types of motor systems

Movement systems (locomotors and contractors)

Locomotor organs are the structures which by leverage and power move an animal through its space. In Metazoa they are primarily muscles attached to rigid articulated skeletons, but there are analogous devices in the Protozoa, some of which have been retained in metazoan animals for slightly different purpses. The main locomotory device in Protozoa are flagellae and cilia, strands of various lengths which generate forces by helicoidal or whiplike movement as a result of the shortening of nine core contractile filaments. The system has been retained in Metazoa largely to circulate internal fluids; the cell stays still and the force generated circulates the fluid. Other locomotory systems include the transitory organelles of pseudopodia as cytoplasm is powered forward against the surface tension layer of the body cell in Amoeba and related forms.

Among the Metazoa, muscular systems become increasingly more complex with body size and mass, especially once animals emerged from water, became terrestrial, and needed to power their masses through the less buoyant medium of air. Thus two-dimensional animals such as flatworms could glide over surfaces, or burrow, by a simple muscular arrangement generating wave action along the body and transferring forces en route. Coelenterate plankton could function in the marine water column by contractions of the radial musculature of the bell. But on land little could be achieved until body cavities, segmentation, and skeletons evolved in arthropods and vertebrates. These major evolutionary developments allowed the convergent shaping of body forms with powerful leverage in the two main lines of descent for complex animals. The potential is enormous and is applicable to almost all essential functions from feeding and mating to an infinite array of counter strategies between predator-prey, competitors and cooperators.

Contractor systems deploy muscles or alternative organs in a different way. Contractions serve to change body shape, usually to make the body or parts of it smaller without moving the animal from one place to another. For sessile forms this is the limit of their movement. The contractions of stentors (protozoans), sea anemones, and bryozoan zooids represent their escape responses to ecosystem disturbances, and there is little else that they can do, although the metazoans may expel gametes and secretions by contractions as well. Sedentary and even quite active animals may also use contraction in various ways. It may be an escape response to predators or disturbing stimulii, especially when the animal is provided with a hiding type of shell (snails, clams). It may serve to bring in exposed body parts, as with the hanging manubria of some drifting coelenterates. But contraction (with the associated expansion of relaxed opposing muscles) may serve to shape the body of a complex animal into a postural display, such as the chest inflations of the courting male sage grouse.

Call systems

Calling organs produce sound or transmit ultralow and ultrahigh mechanical energy. Humans are sensitive to mechanical energy frequencies from 20-20,000 Hz, and we call this window "sound". We hear sound, primarily through external collection by our ears, although recent bone radios, draped over joggers' shoulders, draw on the ability for sound conduction through the body. Like many other animals we transmit sound shaped into calls with a variety of informational meanings. In our case most of our calls are so complex that we designate them as language, but there are simpler emotive screams, gasps and giggles analogous to the calls of other animals. Animal calls have evolved as adaptations for alerting kin, luring mates and prey, maintaining contact with kin and mate, and threatening competitors. There are common patterns in these uses. Alert calls tend to be short, noisy and directionless, so that the caller does not give away his or her position. They may be prey-specific. This is a forerunner of the complexity of transmissable information between calling conspecifics. This complexity ranges from the limited variety of calls in even socially organized chimpanzees, to the almost infinite variability of human conversation.

Calling includes transmitting outside human sensitivity, at both lower and higher frequency. In fact, our low frequency levels are just about as good as any other animal's although we may lose some of the vibrational components found in quail drumming, lion purring, and similar throbbing calls. We can also hear underwater the far ranging calls of whales, some of which at 20 Hz potentially may travel almost around the world through a deep ocean sound window, and thus serve as a social signal between remote individuals in a medium which is opaque to light.

Some high level calls may be just beyond our frequency range, as are various whistles and chirps of insects, but there are various animals which transmit ultrahigh frequency sounds equivalent to the sonar which we artificially transmit for sea navigation. Bats are prime users of this system. They pulse-transmit several different calls, the echoes of which serve the function of showing the configuration of space ahead of them, including the location of flying insect food. The evolutionary arms race shows well in this example. Some moth prey species have evolved a counter adaptation of immediate freefall on receipt of the bat sonar transmission, so they survive, and the bats take other moths lacking the adaptation.

Smell systems

Many animals produce pheromones from secretory glands. These may serve the function of releasing mating behavior, as in many moths and butterflies, or marking territories in social mammals. However, pheromones oozing from glands are barely functional unless provided with a long distance transmitting system. A few species secrete pheromones which are sensed by close sniffing contact between conspecifics, but the majority have evolved physical transmitting methods. Commonly, the pheromones are vaporizable, disperse to ambient air, and as they drift downwind can be received by target individuals. This is the common procedure in insects. Aquatic animals, with non-vaporizable pheromones, use essentially the same process, often for synchronizing gamete release. Bivalve molluscs, which lack intromittent fertilizing organs, are an example. Yet other terrestrial species physically transfer liquids containing pheromones by squirting or rubbing behavior. In the well-known case of dog territory marking, the pheromone is included in urine, but in other species pheromone-specific glands may have an associated muscular system for jetting the fluid at habitat or conspecific.

Variants of smell producing organs include those which produce defensive secretions as in the jets of skunk, and many other less well-known animals such as the fulmar and other spitters.

Pheromone production generates an arms race type of problem between predator and prey. Either can release smells for their own purposes, but the other has the advantage in detecting that smell for their own survival. Accordingly, active predators commonly stalk from downwind, and sharks may track prey in still water in a manner strongly suggesting that a detectable plume of chemical had been left behind. The pheromone arms race between people and insect competitors has reached the level of synthetic mass production for systematic application and research.

Light systems

Marine bioluminescence is a fairly common and well-known phenomenon. Some animals, however, especially terrestrial forms, have evolved methods of transmitting light patterns as a source of information about themselves. The light patterns may be relatively simple, barely involving behavior, as are the illuminated lures of abyssal angler fish, or the still patterns shown by stationary glowworms. Some fireflies have introduced significant behavioral extensions to their potential for information transmission, so that carried lights are flown in distinctive species-specific ways, or flashed in distinctive call signs like those of lighthouses on dangerous coasts. In some cases grouped fireflies can flash in unison, escalating the visibility of their species-specific signal.

Electrical systems

A few fish species have adopted into their transmission systems low powered cellular electrical phenomena, to which humans are sensitive only through instruments detecting brain (EEGs) and cardiac (ECGs) waves. The South American turbid water fish *Eigenmannia* generates an electrical field between its mouth and body which allows the detection of electrically opaque animate and inanimate objects when they are near enough to shade the receptors. Electric eels and rays have adapted muscle cells to generate, store and transmit lethal electric shocks at prey or predators. Even people are sensitive to the shocks of electric eels without using instruments.

The possibility of presently unknown systems

We know that in the environment energy forms exist to which some animals are sensitive and can respond adaptively. We also know that only some of these energy forms have been drawn upon by animals to evolve transmitting systems which serve to manipulate other animals. These energy forms are light, sound, taste-smell and electricity. What about non-visible electromagnetic waves and magnetism? Are there animals with transmitting capability for these energy forms? We should remain open to this possibility as well as the possibility that there are other energy forms to which humans are not sensitive but to which some animals are. In short, there may be energy transmitting systems that have not yet been discovered. But it is easy to be uncritical of hoaxes and nonsense in this field of ESP, parapsychology and morphogenetic fields especially where the scientist has high credibility in another legitimate discipline. Gardner (1981) claims that testing of media, ESPers, and similar people should have experienced professional magicians as part of the investigating team to protect honest naieve scientists against cheating by human subjects. The physician Arthur Conan Doyle, for example, created the rational detective fiction of Sherlock Holmes; but he also believed in fairies, and had photographs to prove they existed. The physicist Einstein once wrote a preface for a book *Mental Radio* but later concluded that the absence of distance effects in ESP indicated a "non-recognized source of systematic errors".

Motor system quality control

Animal behavior can be well directed towards or away from sources of stimulii, or even tangential to them. We can look on this directional capability as a form of quality control. Behaviors may be produced as a response to a stimulus, and need no directional component. The simultaneous spawning of clams and oysters in a drifting medium, for example, draws on the directionality of the current to disperse and mix gametes. However, the mating of two internally fertilizing arthropods or mammals requires precise directing and timing. This is achieved in two ways: internal and external controls.

Internal quality control of direction and timing is an interface phenomenon between motor organs and nerve system. The need is met by an internal sensory system, the proprioceptors, which repeatedly feed back to the coordinating center information on the state of the organ. The behaving animal is thus able to monitor its performance, collating information derived from the external environment and making adjustments if necessary. There is some possibility that this sensitivity has been escalated into an Inertial Guidance System for long-distance movements, by which a migrant can keep track of distance covered and courses set and hence be able to locate itself on a map held within its brain (see Chapter 8).

Quality control of performance is also achieved by monitoring responses of other animals affected by the performance, or by the effect of the response on passive ecosystem components. Thus a stickleback building a nest, or a greylag goose rolling a displaced egg back into the nest, directs its attention to the product of the behaviors. It receives information, and modifies or stops its actions as appropriate. The greylag goose and its egg retrieval behavior shows some limits on the quality control system. The goose can only counter sideways egg rolling slightly because if the egg rolls away (as ovoid eggs tend to do) the retrieval action has to be completed before the goose can restretch its neck and begin again.

The stickleback, after building its nest, has the more complex quality control problem of directing its behavior to a responding other animal. It directs its attention, i.e. it deploys relevant sensors (eyes and lateral line), to the courted female so as to produce appropriate behaviors not just in the right sequence but at the right time to cross-stimulate the female.

The other interface phenomenon of how the actions of the motor organs are initiated by nervous control is a matter of physiology, and can be black boxed.

Coordinating systems

All behaving animals have to process information about the ecosystem in such a way that they respond adaptively with their motor organs. They do so with a system of hardware which allows transmission of information about the received signs and configurations from sensors to motor organs. The hardware also allows the processing of the information into a form which is meaningful to the animal. The animal responds in an adaptive way. It adds to its chances of staying alive and breeding.

The following comments provide a very brief summary of anatomy and physiology for the context of understanding the nature of the black box in the behaving animal model (Figure 1.5). The summary needs supplementary reference to standard texts on animal physiology, preferably source books on comparative physiology (e.g. Schmidt-Nielsen 1979).

Structural and physiological basis

Among Metazoa the nervous system is the best known form of a variety of information transmission and coordination systems, but there are other analogous systems which have been given the general term "neuroid" (Mackie 1970). The nervous system works on the basis of a wave of electrical depolarization travelling along an elongated cell axon, followed by a wave of repolarization. The minute gap (synapse) between cell extensions and the next cell are leapt by a squirt of one or other transmitter chemicals. The system moves information around quickly. If you are a small animal, and the information does not have far to travel, you can be active. A low speed of transmission, however, means that a large animal will inevitably be slower in responding than a small animal, especially if it learns a lot and has to retrieve stored information for integration with the new.

Large active animals have converged on rapid transmission systems using two tactics. The tactic of the Protostome line of descent has been to extend nerve cells to relatively giant size so as to minimize the number of synapses, at which transmission is slower than within the nerve cell. Annelids, arthropods and cephalopods all have these giant nerve cells, with axons bundled together like electrical wiring connecting ganglia (brain) and muscles operating quickly for escape or attack.

The tactic of the Deuterostome line of descent has been to increase the speed of transmission within the nerve by insulating the axons (a myelinated layer). This generalized approach allows rapid transmission throughout the body, leading to overall quick coordination. The limits of the system appear with animals the size of the larger dinosaurs and whales. The dinosaurs may have shared with whales the assist of a raised and controlled body temperature (Desmond

1975).

The neuroid systems come in a variety of forms and transmission speeds (see Figure 2.2). Those in the coelenterates and tunicates transmit information only slightly less quickly (10^{-1} m/sec) than the slower nervous systems, but sponge systems (there are at least two of different speeds) are slower. Essentially these systems allow information transmission to proceed slowly in animals which are not built for speed and activity, but which nevertheless retain the need for some protective information. They can do something about their environmental situation, such as turning, feeding or contracting external apertures.

Even the Protozoa have converged on the generalized animal resolution to the information transfer-coordination problem. Information transmission in the form of propagated electrical changes occur in the dinoflagellate *Noctiluca* and are associated with tentacle food gathering and bioluminescence.

The chemical fallback resource of the nervous system whereby secreted fluid is physically squirted across the minute distance of the synapse, shows an aspect of the integration of electrical activity with chemical production. This property is greatly amplified in many animals, and there is continuing interaction between the electrical and chemical properties of the transmitting cells of nervous systems. In animals with a fluid transmission system, there is potential for chemicals from nerve cells to be physically transmitted in solution more pervasively, even if more slowly, than by electrical phenomena.

This chemical alternative transmission system has been specialized in active animals, especially arthropods and vertebrates, and deployed in the form of distinctive tissues or organs which produce one or more endocrines, which can provide short or long-term pervasive controls on the body and its activities. Endocrines can prime the body for certain activities.

The basis of information transmission then, is a structure of transmission lines ramified through the body, but concentrated into nodules (ganglia, brains) for coordination functions. The transmission lines are capable of integration with the production of chemicals, a property which also may have been modularized into distinctive chemical producing organs.

The concept of priming

Many animals only respond to stimulii at certain times in their lives. It is adaptive to do so in the sense that they do not waste energy responding like automata to environmental situations which are not presenting survival problems at that time. The mechanism is termed priming. If the animal can be primed only at certain times when the stimulus complex is most likely to be important, then its responses can be limited to those times.

An example is the seasonal sexual maturation of many animals which live in temperate and polar regions where summer is the biologically productive time, thus facilitating feeding by a new generation of voracious and growing young. For many birds and mammals which overwinter in temperate and subtropical regions, many of the habitat and biological components of the ecosystem remain satisfactory for breeding, but others do not. The balance is not as satisfactory for successful breeding as it is in summer further north. Accordingly, a physiological system which primes the animal to respond seasonally, is a relatively simple way of adapting to the situation.

Many other complex behaviors can be brought under a priming mechanism, either regularly or once in a lifetime. Thus insect diapause involves complex interplays of hormones, and there is some indication of similar hormonal controls in the spatfall and final habitat selection of pelagic larvae. Priming is a tactic for achieving the strategy of responsive periods. It was originally modelled conceptually by Lorenz in his flush toilet model (Figure 1.1) and by Tinbergen in his "wiring" diagram. The latter shows some of the essential features of priming, and is worth detailed review (Figure 7.1) any time controlled activity can be considered as a hierarchically arranged sequence. Only when an animal is primed for such activities as breeding will it respond to appropriate stimulii directing it to take the first behavioral steps in the sequence. In the stickleback, this involves migration to the breeding habitat. Once it is there, later steps of nest building, courtship and mating can proceed in sequence. The stickleback remains primed by its internal flooding of sexual hormones, but as the breeding season ends so the hormonal priming disappears, and the fish reverts to its vegetative activities. The priming cycles have been described in other ways (Figure 7.2) and other fish, and many physiological texts present the endocrine basis for estrus in mammals, the metamorphosis of insects, and the behaviors associated with both.

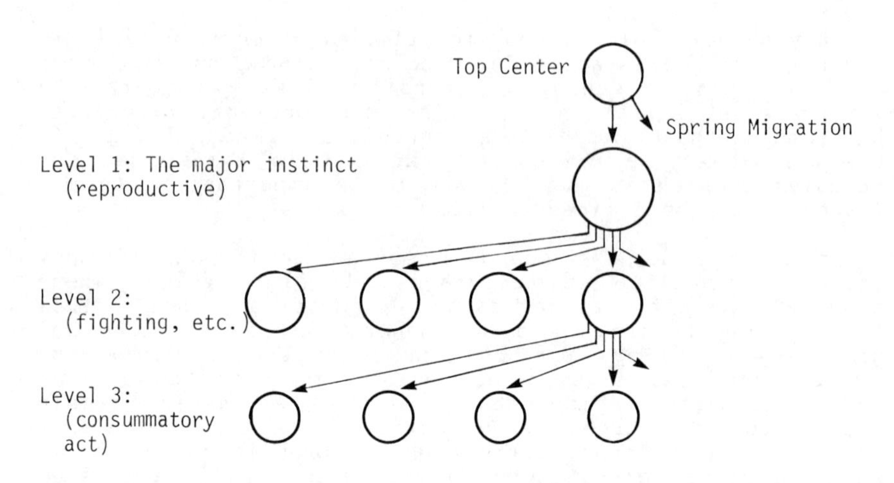

a. Overall Model

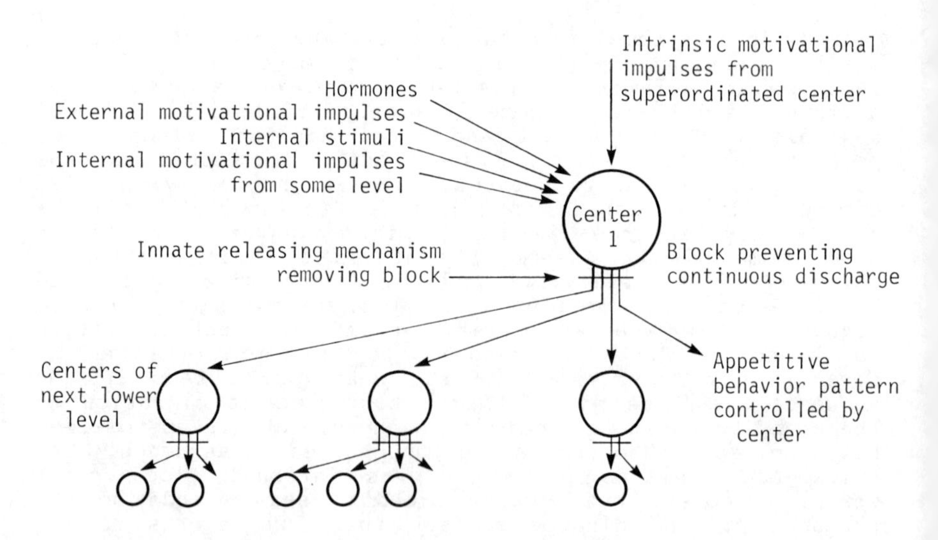

b. Priming Mechanisms

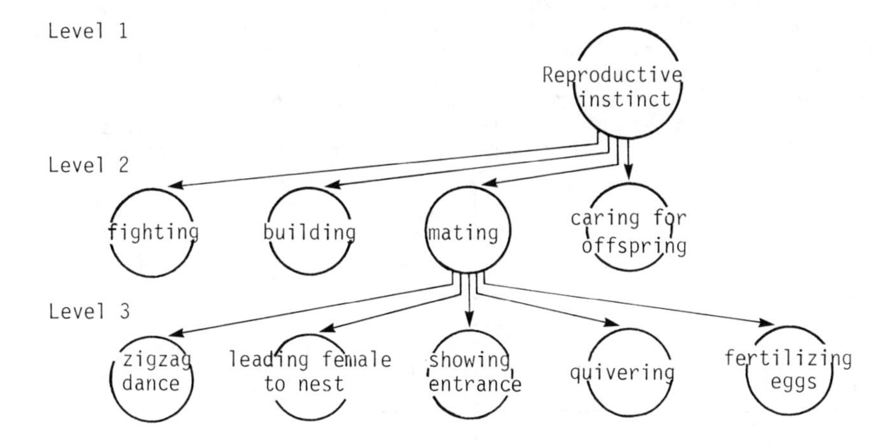

c. Behaviors primed at various levels

Figure 7.1. Tinbergen's (1951) model of hierarchically arranged priming and responsive periods for the stickleback. A coordinating mechanism (.e.g. top center in a or center 1 in b) is activated and progressively other lower level centers are primed (a,b and c).

Lifetime effects

During ontogeny there can be critical periods of cell, tissue and organ differentiation which at least in part are under chemical controls. The tactic for achieving the strategy of a consistent species anatomy and physiology while allowing for intraspecific differences, such as male and female sex, is to retain a species gene complement but to control its expression in some way. Thus sexual differentiation in two-sex species is usually determined at a critical early growth stage when sexual characters can still be shaped. The sex of mammals is determined during early pregnancy, and only occasionally goes wrong to the extent that intermediates are formed (and in humans need hormonal or surgical treatment). The system is sufficiently flexible in timing to accomodate the tactic of social organization in some cleaner fish, whereby on the loss of the dominant male the top ranked female changes sex and becomes dominant to her (or is it his?) social group (Robertson 1972).

The tactic is believed to function in the determination of many intraspecifically variable characters during embryogenesis and can be traced back to the ultimate chemical controller of body form and function, i.e. DNA. Once again the behaviorist is at a stage of deciding at what level of understanding, in this case embryogenesis and ontogeny, he or she must revert to black boxing the

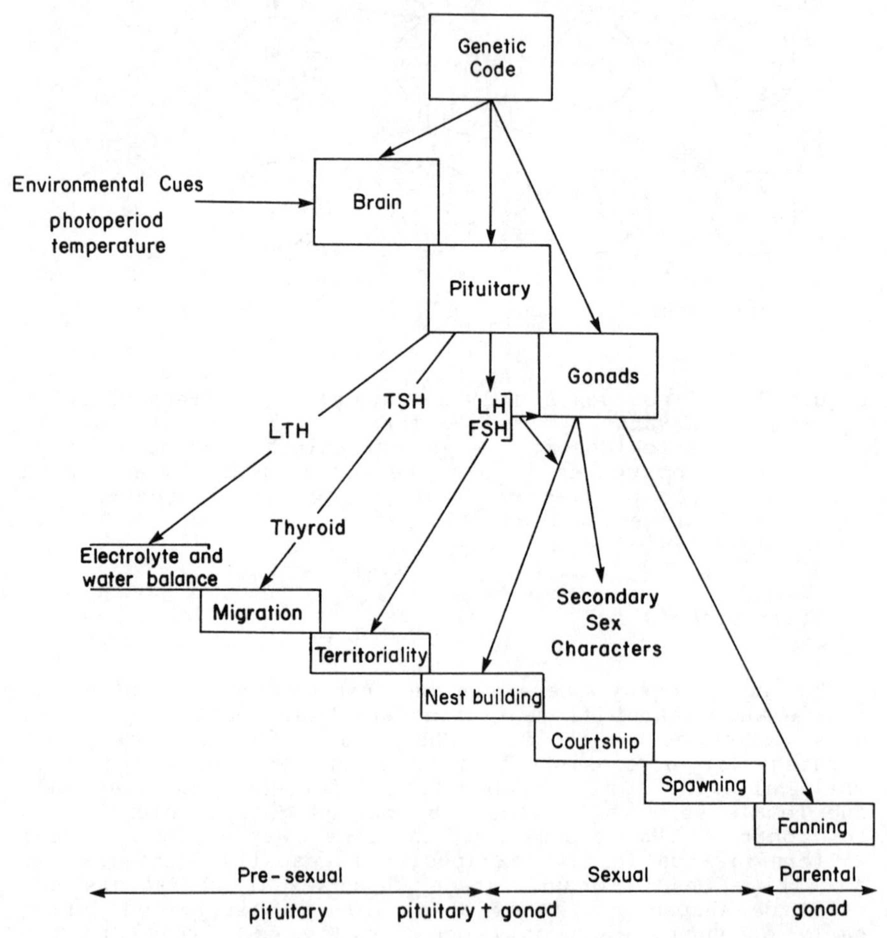

Figure 7.2. Priming diagram for sexual maturation in stickleback (modified from Hoar, 1965).

causative mechanisms.

Coordinating centers

The information processing systems of complex animals have a discrete coordination center (a brain) which is a tangle of transmission lines somehow organized to produce adaptive behavior. A brain receives new information and collates it with information already in storage, prior to sending instructions to selected motor organs so that they respond effectively. This means, in turn, that the brain must have data storage, retrieval and comparison facilities as well as coordinative circuitry. In practice the mammalian brain appears to store information at a variety of locations, such that the representations of the outside world via various sensing systems drawing on different energy forms can be stored separately and diffusely. The system is almost unimaginably complex. It has been accommodated by retrofitting extra forebrain capacity of R-complex, limbic complex and neocortex on to a vertebrate neural chassis of mid-brain, hind-brain and spine.

The brains and ganglia of simpler animals are more amenable to analysis. The coordinative capability of some species with a known, limited number of neural cells has been analyzed to a level that the cell interlinks can be described like a wiring diagram (Alkon 1983) and learning processes demonstrated in terms of modified nerve excitability.

The biologist interested in overt activities of behavior rather than physiology needs to understand the capability of the brain in terms sufficient to meet research objectives. The extent to which the behaviorist needs to understand the details of anatomy, physiology and localization of brain phenomena depends on individual interest, the species concerned, and the nature of the behavioral phenomena that are being investigated.

Summary

When we understand *how* an animal's behavior is caused in terms of its physiology and ontogeny we understand proximal causes of behavior. But we can also understand *why* it behaves the way it does, rather than in some other way. This latter means understanding how natural selection provides the ultimate causes for particular behaviors.

Proximal causes for behavior include the mechanisms of sensory perception, information transmission through the animal body, coordination of information from various senses and from prior storage, communication of instructions to motor organs, and the physiology of motor organ activity.

Proximal causation also includes understanding molecular controls on these organ systems in development based on coded instructions in the animal's gene composition.

The behavioral biologist needs to understand the potential capability of each of these systems. The levels at which the details must be understood can vary considerably. Individual biologists must decide what level of understanding of the mechanisms of proximal causation they need for their particular behavioral objectives.

CHAPTER 8. LONG AND SHORT DISTANCE MOVEMENT.

When an animal moves it may be just for a short distance as when it attacks something else, flees, or simply patrols its territory. The distance is short in relation to the animal's size, its speed, and the duration of its activity. Essentially it is responding to stimulii which are signs for environmental conditions immediately around it. The sign stimulii originate from the environmental conditions to which the behavior is adaptive. These were the only effective stimulii for small primitive animals incapable of travelling far relative to the scale of environmental gradients, and remain the only stimulii for present day small animals unless adapted to utilizing environmental drift in air or water.

However, there are occasions when larger animals initiate a very different type of movement. This is when they start moving in a sequence which will take them far from their starting place. An example is the seasonal migration of birds. "Far" means not only a distance enormous relative to the animal's size but enormous relative to the animal's speed, so that the duration of the movement is greatly extended over normal activity periods. There are obviously a number of significant behavioral phenomena and problems here.

On long-distance movements, particularly at starting time, is the animal receiving sign stimulii from the remote environmental situation which it will eventually reach? If not, what sequence of stimulii are involved, and how could the animal become programmed to respond to such a sequence of stimulii? Do concepts of long and short distance movements really represent different phenomena or is there no sharp dividing line between the two (Figure 8.1)?

The problems in understanding long-distance movements, like those of other behaviors, reduce to the two categories of proximal and ultimate causes. This chapter will consider mainly the proximal causation of long-distance movements but also associated topics such as periodicity, which are important proximal cause processes. Thus this chapter deals with yet more levels of understanding at the proximal level, levels which are different from those already considered

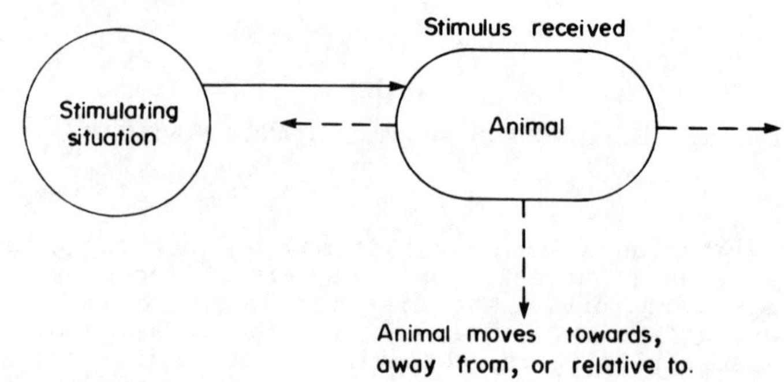

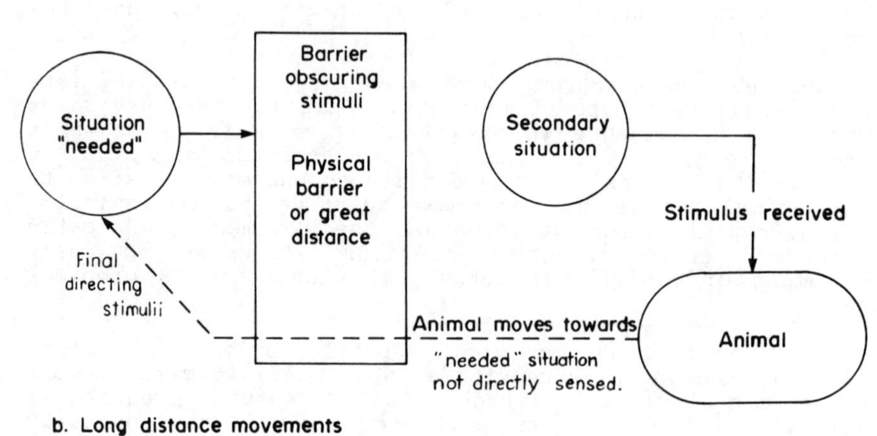

Figure 8.1. Conceptual models of long and short distance movements.

under programming (Chapter 5) and physiology (Chapter 7).

The procedure for understanding causes, as usual, should be in part inductive. Information will be collated, patterns of long-distance movements will be documented, and then deduced causative concepts for the behaviors will be examined.

The phenomenon of long-distance movements is significant for another reason, and that is the extent to which humans impact the behavior of animals. When animals engage in long-distance movements they can range into environmental situations which are considerably affected by human activity and which may either misdirect the movers or cause serious population losses. This matter is taken up in Chapter 9.

The patterns of long-distance movements

Figure 8.2 and Figure 8.3 demonstrate a set of patterns which appear when long-distance movements are considered in terms of a number of contrasting phenomena: active or passive; solitary or social; round trip or one-way; if round trip, one-shot or repeats; if repeats, periodic or irregular; if periodic, daily or lunar or annual or other period; and one generation or several generations for completion. These topics all appear when the variety of long-distance movements is considered, and they have considerable implication for understanding proximal causes.

Figure 8.2 gives only enough examples to illustrate the topics to be raised. Other examples can be added to the boxes.

Some of the best known long-distance movements are one-way as opposed to being round trip. The so-called "migration" of lemmings is one-way. It is a mass movement (Elton 1942) away from an overpopulated center, with the animals in an unusual hyperactive state which drives them to move forward in spite of lakes, cliffs and other obstacles impeding or terminating progress. As the mass movement loses individuals through death or diversions, the hyperactive state disappears, and at least some individuals may have transferred themselves to a new habitable area. It appears to occur only in certain parts of the range of the species, i.e. in Scandinavia and not North America. It is an emigration, a density dependent phenomenon which happens to occur periodically (at four year intervals) due to population cycles within the species. It is a very different phenomenon from the annual round trip migration of the species between winter and summer habitat, in which individuals move solitarily along established trails.

Another one-way movement is also a well-known "migration". The Desert locust (Johnson 1969), in addition to a few species in other parts of the world, is a hyperactive grasshopper which, in overpopulated conditions, develops into a new anatomical-behavioral form and departs

		One-way	Round trip						
			One-shot	Repeats (within one generation)					Several generations to completion
				Irregular	Periodic				
					Daily	Lunar	Annual	Other	
Active	Solitary		Salmon Cicadas	Home rangers	Home rangers, Zooplankton	Tidal invertebrates	Birds (broad front), Lemmings		Monarch butterfly
	Social	Lemmings	Salmon		Home rangers, Fiddler crabs	Fiddler crabs	Birds (Narrow front), Eels (adult), Oceanic vertebrates	Salmon, Turtles Whales	
Passive	Solitary		Salmon		Zooplankton		Birds, Eels (larvae)		
	Social	Locusts	Salmon				Birds		

Figure 8.2. Patterns of long-distance movements, with examples.

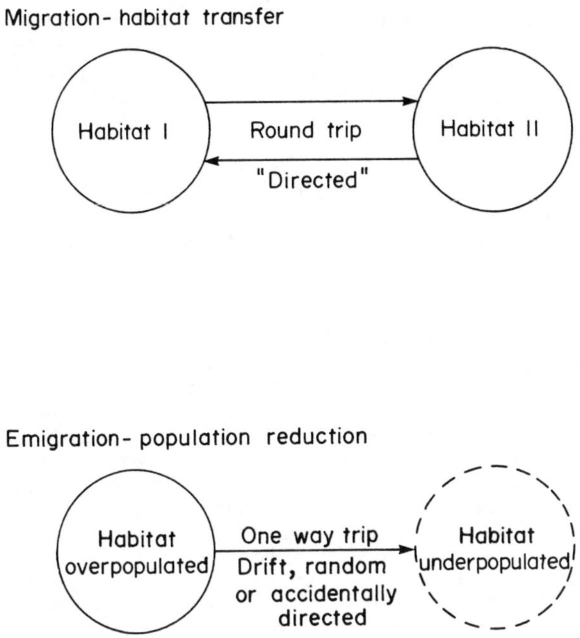

Migration- habitat transfer

Habitat I

Round trip

"Directed"

Habitat II

Emigration- population reduction

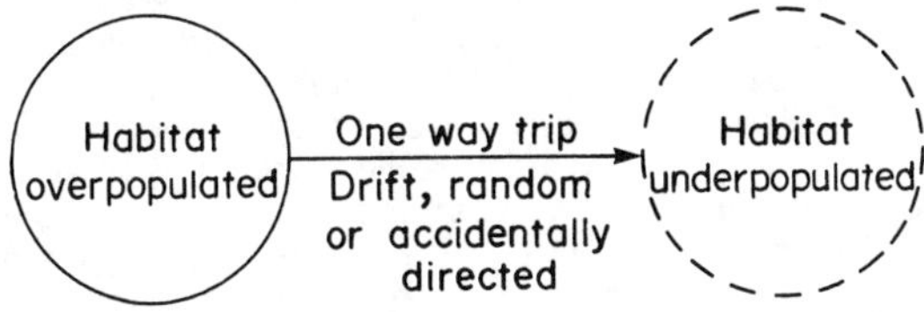

Figure 8.3. Models of long-distance movements illustrating the significant differences in the convergent tactics of long-distance movement attaining the two strategies of habitat transfer and population reduction.

in mass flights from the population center. The locusts differ from lemmings in two ways. They not only change form, their flying may be randomly oriented. The direction taken is set by prevailing winds, not by oriented movements. Thus these emigrations are passive in the sense that the route followed is determined by drift, not by an orientation process of the insects.

Active and passive movements reflect whether the moving animal's activity and sensitivity are responsible for the distance it covers and its direction, or the opposite, i.e.

the achievement is due to drift from wind or currents. Lemming migrations are active in the sense that the animals are moving themselves, although their direction is open in the sense of being away from the population center. Locust migrations are passive in the sense that the route and distance are basically determined by wind drift, although the hyperactive random flying of the insects maintains them aloft, and can be directed to lead them to a feeding landfall when they are on route.

Round trip movement contrasts with one-way emigration. It encompasses a great array of patterns by which animals return to areas which they have previously occupied. This is the best use of the term migration; one-way trips should be termed emigrations.

Round trips may, on occasion, take several generations to complete. The monarch butterfly (Figure 8.2, far right column) is a well-known example of this phenomenon. The fall generation migrates from northern regions such as Canada over many hundred miles to relatively warm winter quarters in California and Mexico. In the spring, several new generations each migrate northwards much shorter distances as their food species the milkweed progressively grows to an edible size. Thus the individuals returning to the northern summer range not only have never been there before, but each generation flies only a part of the total route to be covered. The distance achieved by the butterfly is remarkable physically, but the peculiarity of taking several generations for the trip implies a gene pool capable of directing movements differently in different generations. The proximal directing causes are unknown.

The monarch butterfly achieves its migrations solitarily. Each individual moves independently of the others as far as is known, and hence each is independently sensing appropriate sign stimulii from the environment and directing its movements. The same process largely determines annual lemming migrations, although these species appear to sense by pheromones that other lemmings have used the trails. Some butterflies may have the potential to follow others on migration since follow-my-leader migrations have been recorded with individuals following sufficiently remotely that they were out of sight of each other. Nevertheless, these solitary species are not drawing on the social response of following an individual leader who is personally making decisions on the route to follow. Such leadership behavior has implications for directional sensing in that the sensitivity may need to be developed or deployed well only in the leader, while others take the same route as a result of an alternative proximal option - following a leader.

Migrations are commonly completed within a single generation. Each animal makes the whole trip. However, there are different implications if the migration is made more than once. Animals such as Pacific salmon, which only migrate once, must be capable of directing themselves over all segments of the route, and cannot rely on following a leader who has learned the route previously. Pacific salmon carry this individual capability through a complex sequence of life stages (see Figure A2.1). The alevins hatch about 1 ft deep in gravel and must move up near the water interface so that at the right time in spring they can release themselves to the water column. Species differences determine in part the next stages in the sequence as they migrate downriver to the sea, taking from one night up to two years to do so, some with freshwater residential-territorial stages en route. In the sea they may spend a feeding period in an estuary before departing. At sea they may be coastal, or range into the high seas of the North Pacific. Finally, they return to coastal waters at the right time of year to locate their home river, and complete an up current migration to their home gravel bed before winter storms and freeze-up. (For more details see Appendix 2.) None of the salmon repeat. None can learn the route.

In addition, Pacific salmon have a complex direction-finding strategy of relying on active movement (lake crossing, oceanic departures and return, and upriver swimming) and drift (downstream migration and high seas gyral).

In many migratory species individuals can repeat the trip several times, and the potential to learn routes is a tactic which can raise the benefit-cost ratio of the strategy for changing habitats. Within such repeating species the frequency of the trips varies.

A few species show irregular movements, generally covering relatively short distances. We often do not consider these as migrations. This category includes home ranging by groups of chimpanzees, elephants, and other species. They have large ranges which are only patrolled irregularly, perhaps for specific reasons such as finding certain foods, or water in a drought. Sign stimulii originating at one end of the range are too far away to be received at the other end. The common property that these animals have is a complex brain with memory power sufficient to learn routes and locations or range components. In behavioral terms home ranging, whether irregular or periodic, is a long-distance movement reduced to short-term behavior through the learning ability of the species and social living. Novices can follow the experienced and learn routes and timing so that the movement is only of short duration.

Periodic migrants show a scale of time patterns. There are daily, lunar, annual and a few long-term cycles.

There is an enormous array of daily migrants in the sea as zooplankton and some of their predatory nekton move up and down in phase with reducing illumination or other sign stimulii of diurnal changes (Longhurst 1976). The behavioral strategies vary and may be complex. Movement can be active swimming, a passive adjustment of buoyancy achieved by the production of gas bubbles or oil droplets, or a mixture of swimming one way and floating or sinking the other.

In habitats like the shoreline where the environment is under various daily and lunar physical forces, migrations can follow the physical changes. Thus many transient bird and mammal predators visit beaches at low tide, and fish predators may range the same habitat at high tide. Some daily rangers may adapt their movements to an interacting lunar rhythm. Fiddler crabs only emerge from burrows to the dangerous but food providing beach habitat when low tide occurs at night and they can both feed on beach organics and court mates. Their tidal rhythm continues if they are transplanted to laboratory conditions. (Palmer 1975).

Annual migrations are commonly experienced in temperate and polar regions because summer conditions provide abundant insect and plant food supplies on which young can be raised. Many bird species fly north-south routes over large distances such as between North and South America, or Europe and Africa. A few have even longer and directionally more complex movements, such as the arctic tern and its circum-Atlantic, even global migrations between the Arctic and Antarctic. Their strategy may be to utilize an assist from favorable winds, but essentially they are active migrants. The longest recorded journeys are of the order of halfway around the world. For instance, an arctic tern *Sterna paradisaea* was banded in the Farne Islands, U.K. as a chick on June 25, 1982, and retrieved 115 days later in Melbourne, Australia. Since it left the coast in late summer, it must have flown about 100-200 miles a day for about 60 days. In May of 1956, another tern was retrieved near Fremantle, western Australia, after having been banded in the White Sea in July 1955.

There are two other contrasting strategies in active migrations (Figure 8.4). There are broad front migrations in which individuals travel more or less independently in the right direction, maintaining contact by calls if at all. Many small passerine species broad front migrate over flat land and seas and even (if flying high enough) over mountain ranges at night. They lift off from and descend to daytime rests after dusk and before dawn respectively, thus minimizing predation as they arrive tired from extended

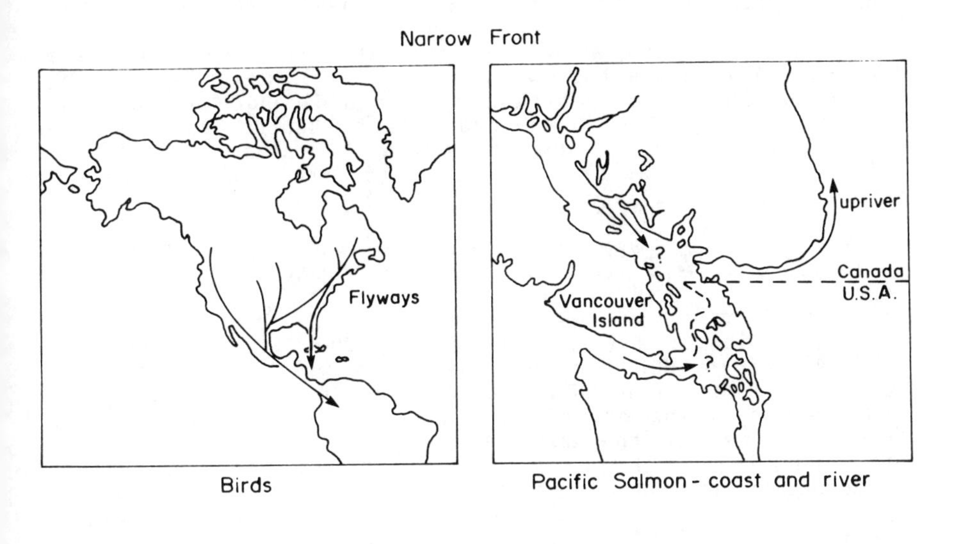

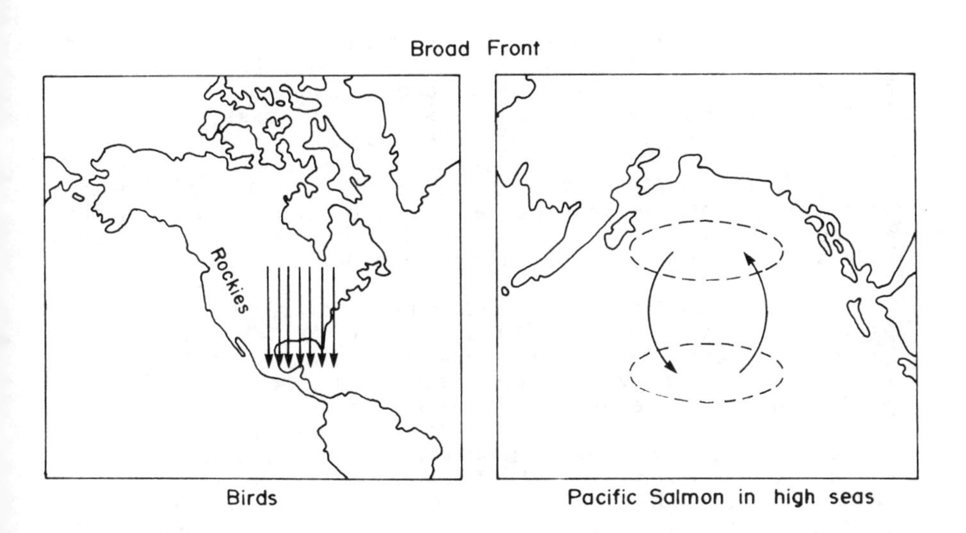

Figure 8.4. Narrow and broad front migrations among birds and fish.

flights. They are subject to a number of environmental dangers, from wind drift to loss of sensed direction.

Narrow front migrants, in contrast, are largely daytime movers and visually follow environmental guidelines (leading lines) which include coastlines, mountain ranges at the edge of flat land, and river channels. Routes include the waterfowl flyways of North America -- Pacific, Mountain, Mississipian, and Atlantic -- and the equivalents in Europe/Africa and Asia.

A few other special migrants must be mentioned to illustrate some of the more complex strategies for habitat transfer.

Eels spawn in the sea at depth. The larvae drift to a coastline providing river habitat in which the juveniles will grow. On maturity the adults return to the ocean, swim to their spawning ground (which may involve a transoceanic crossing, as for the European eel), and descend to depths for spawning. The strategy, as in some other species, combines drift and active components, and its ultimate and proximal causes must be understood in terms of stimulii initially provided by prehistoric geography but changing over evolutionary time.

Many marine vertebrates make extensive migrations over or in featureless oceans. Commercially fished species such as herring and anchovies appear and disappear annually on fishing grounds. Humpback whales annually pass from Hawaii to Alaskan fjords, and along other routes. Sea birds range over great areas, returning to home sites to breed annually or at longer intervals. Birds have some capability for landmark orientation as they are flying high towards islands or continents, but orienting stimulii for long-distance underwater marine migrations remain a puzzle.

Turtles make similar submarine migrations over long periods of several years (Bjorndal 1981). They leave beach breeding grounds when barely able to scamper into the sea (a well directed active movement at night with a minimum of predators) and return many years later to their home beaches. These may be on islands (such as Ascension Island in the Atlantic), thus requiring considerable feats of navigation. Ascension Island is only five miles wide and 1400 miles from Brazil. It represents the last of a chain of sea mounts created by continental drift as Gondwanaland split into South America and South Africa (Figure 11.2). The long migration may have arisen by gradual extension of the breeding base from one sea mount to another over 100,000,000 years.

Finally, there are rare long-term one-shot or periodic movements. There are a number of species of cicadas in which sexually mature adults emerge from the ground to fly and mate briefly but at long intervals ranging up to 17 years. There is no known corresponding environmental cycle. Approximately a quarter of homing Atlantic salmon return downriver as kelts after spawning, and if they survive more years in the sea will migrate upriver again.

In summary, many animals have converged on long-distance habitat transfer strategies in their balance for survival, and they have done so using many different tactics. In the really long-distance cases, understanding the phenomena requires relating current proximal causes in terms of the stimulii involved to ultimate causes, i.e. prehistoric geography, the evolution of extended sensitivity and motor behaviors facilitating sensing over long distances, and maintaining direction.

Causative concepts

Taxes/kineses/tropisms

The term taxis has suffered under different meanings since it came into general use during the 1920s and 1930s. In the 1950s the school of ethology took it over and applied it to the directional component of FAPs. Thus when an FAP was released by a sign stimulus, other stimulii could still affect the action by directing it. These were called taxes. Hunting animals closed in on prey through the direction of taxes once hunting behavior and particular attacks had been released. Taxes directed the greylag goose to stretch out its neck in the right direction towards a displaced egg. But taxes had their limits and if the egg rolled away on its ovoid surface after minor corrective adjustments of the goose's beak, major changes of beak orientation were not possible. This concept has some value since it directed the attention of causative analysts to the possibility that the stimulii directing a movement may be different from the stimulii releasing it, but both are part of the proximal causes.

The long-standing meaning for taxis as a response to a stimulus was developed in detail by Fraenkel and Gunn (1940, 1961; but due to wartime shipping losses sinking most copies of the original edition only the 1961 reprint is usually available in North America). A taxis is a directed orientation response. (I have changed Fraenkel and Gunn's "reaction" to "response" throughout.) Orientation responses guide an animal not only in its behavior, but also in its stance. Taxes keep animals the right way up as well as directing them in what they do. The orientation response is directed, and is not just the leverage of physical forces, for example, crowding clusters of animals into eddies of favorable conditions.

There are three directions to taxes (Figure 8.1a): orientation towards a stimulus (positive taxis), orientation away from a stimulus (negative taxis), and orientation at an angle to a stimulus (transverse orientation).

Positive and negative taxes appear in response to a variety of stimulii such as light and current. It is common to prefix the taxis by a descriptive term (Table 8.1) which, if further preceded by the adjectives positive or negative, provides a description of orientation taken to a particular stimulus. Thus a moth flying into a candle flame can be termed positively phototaxic. The terms are useful descriptors. They also provide some guidance to mechanisms causing the behavior. Using experimental approaches Fraenkel and Gunn (1961) showed that various mechanisms were at work (Figure 8.5). The orientation could arise through the comparison of stimulii arriving on paired or single sense organs which are moving relative to the body. Fraenkel and Gunn called this klinotaxis. The orientation could arise through the simultaneous comparison of stimulii on paired sense organs. If one of the paired organs was destroyed, the animal would break into circular movements. These are the circus movements of Fraenkel and Gunn's tropotaxis. The orientation could be towards a stimulus as though it were a goal. Paired identical stimulii could induce zigzag behavior until the animal made up its mind (telotaxis).

Fraenkel and Gunn developed their theme and produced a large number of descriptive terms, the merit of which is that they reflected the existence of many different orientation mechanisms, and directed attention to their analysis. Today we need to know only the bare bones of the terminology, and that the terms are descriptors. When there is a clear directed response to a single energy form we may be justified in using a term such as positive rheotaxis for up-current swimming, but we should be cautious in applying that term to the upriver migration of salmon. It is almost certainly not just a positive rheotaxis. Undoubtedly salmon swim against the current much of the time or rest head into the current on the way, but they do not do so all the time, and there are many reverse down-current periods of long and short duration. Furthermore, salmon can migrate upriver by choosing energy assisting back flowing eddies, i.e. actually swimming down current in the eddy (see Figure A2.2).

The caution needed in using a descriptive terminology can be pointed out by one of Fraenkel and Gunn's transverse orientations, the dorsal light response. Most animals have a right way up. Usually, what anatomists call the dorsal surface is upwards and what they call the ventral surface is

Table 8.1. Terminology for taxes including transverse orientation (as developed by Frankel and Gunn 1961).

Taxes - orientation of the body of the animal in line with the direction of the stimulus. Locomotion need not occur.

Positive taxis - movement or direction towards the source of the stimulus.

Negative taxis - movement or direction away from the source of the stimulus.

Phototaxis - movement directed by light.
Thermotaxis - movement directed by heat.
Geotaxis - movement directed by gravity.
Magnetotaxis - movement directed by magnetism.
Electrotaxis - movement directed by electrical fields.
Chemotaxis - movement directed by chemicals.
Phonotaxis - movement directed by sound.

Klinotaxis - orientation arising through comparison of stimulii arriving at different sides of the body. Can be mediated through a single sense organ moved from side to side.

Tropotaxis - orientation arising from simultaneous comparison of stimulii arriving at paired sense organs.

Telotaxis - orientation to a stimulus as though it were a goal. Preliminary zigzag movements may occur to paired stimulii.

The three sets of terms can be combined, e.g. a negative photoklinotaxis, as shown by dark seeking maggots.

Transverse orientations - orientation at an angle to the direction of the stimulus. Locomotion need not occur, and if it does is not directly towards or away from the direction of the stimulus.

Light compass response (Menotaxis if locomotion occurs) - orientation at a fixed angle to light rays. This may be temporary, or adjustable with time to maintain a constant direction or reach a goal.

Dorsal (or ventral) light response - orientation so that light is kept perpendicular to the normal body axis.

Ventral earth (transverse gravity) response - orientation so that gravitational force acts perpendicularly to the normal body axis.

downwards. Fraenkel and Gunn showed that the orientation was to light from above (dorsal light response), backed up by gravity (ventral gravity response). When the inner ear of a fish was removed it could be induced to swim upside down by placing a light source below it (Figure 8.6). However, right way up orientation is not just a dorsal light response, and we can suspect that much behavior, especially over long distances, is not just a taxis to a single stimulus, even though taxes may be components of the stimulating situation.

Fraenkel and Gunn developed two other main orientation categories from prior use. The light compass response was applied to animals which maintain an angle to the sun. This forerunner of understanding compass responses to a variety of stimulii exposed the need for a biological clock inside animals since the sun changes position. If the sun is to be used as a compass, the orientating animal needs to be sensitive to where the sun is relative to the stable world around it, i.e. to be sensitive to the time of day. The other orientation category was that of kinesis, which was applied to apparent orientation (aggregation) arising from changes in speed or direction of locomotion according to the presence, absence or strength of stimulii. Clusters of animals can result and such a response is adequate for the accumulation of simple animals in optimal environmental conditions or avoidance of harmful areas, as though they had been directed in their movements (Figure 8.5).

Finally, Fraenkel and Gunn recommended that the term tropism be applied only to plants and not to animals since it had come to mean the specific bending orientations of plants to stimulii during growth. These are mediated by stimulii more persistent than those to which animals have adapted, even sessile plant-like Cnidaria, and Bryozoa.

Kineses and taxes also have phylogenetic significance. In essence, kineses are available to all heterotrophs, including the simplest protozoa since even animals without directional ability can be aggregated in favorable conditions or away from the harmful. Taxes are only available to an animal that can receive directional information from incoming stimulii. This requires specialized sense organs (or organelles in some protozoa). Sequenced taxes (as needed in long-distance movements) are only available to animals with a coordinative system programmed for changes en route. An advanced coordinative system is also need for directed responses to configurational stimulii (star navigation and landscape orientation) and redundancy (multiple backup sensitivities providing similar directions).

These general comments about taxes, orientation and biological clocks now allow us to consider the special

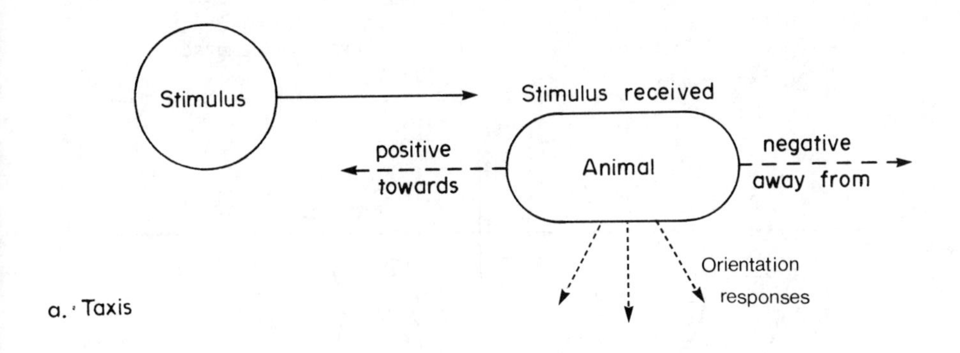

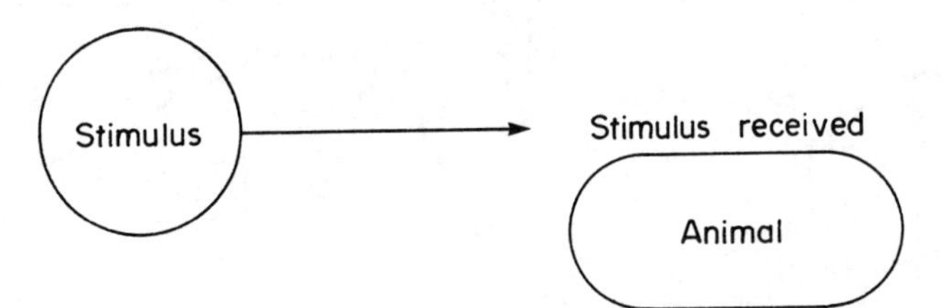

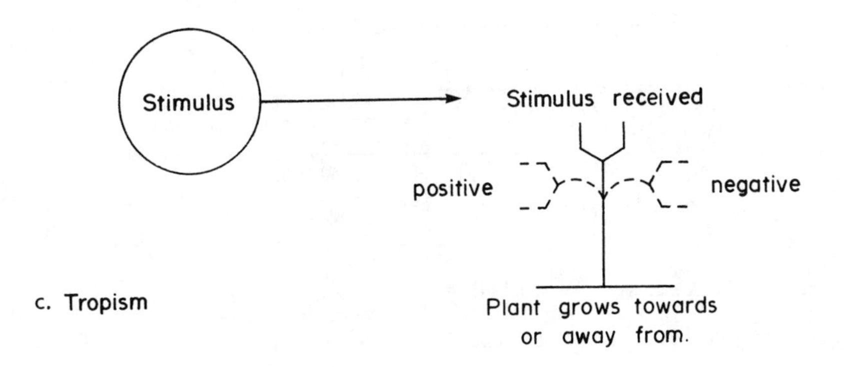

Figure 8.5. Taxes, kineses and tropisms.

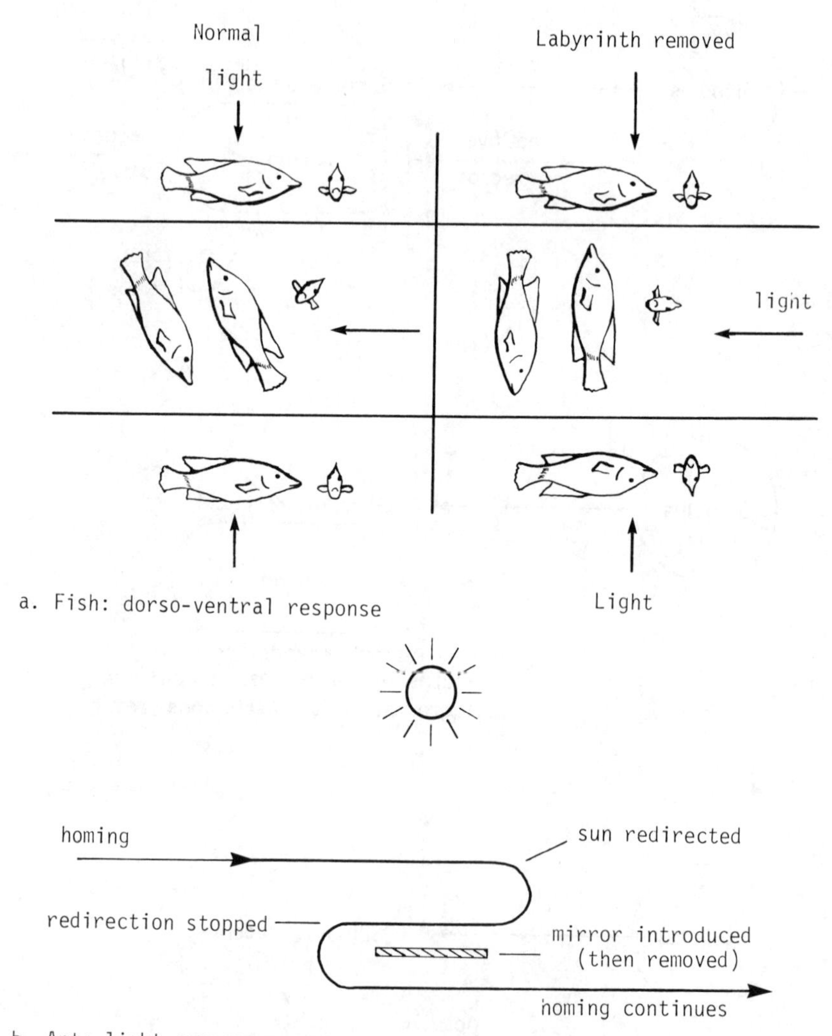

Figure 8.6. Transverse orientations.

problems of long-distance movements to points beyond the range of stimulii emanating from the local environmental situation.

Orientation, navigation and pilotage

Two concepts used in the context of short-term movements are useful entries into understanding the orientation mechanisms for long-distance movements.

The first concept is that of configurational stimulii (Figure 8.7). It is conceivable that long-distance movers can respond to remembered landscape patterns sensed visually either when on their way (leading lines - a coastline for example) or as they arrive at their destination, especially if the movers travel in flocks of mixed repeaters and novices. Among marine mammals and fish migrants this can be called pilotage. It is also conceivable that aerial migrants might maintain a compass direction to landscape patterns sensed visually since mountains and coasts have sufficient geological durability to provide consistent guidance during evolutionary changes. Whale sonic scanning might also permit this. It has even been postulated that the compass direction could be detected remotely from such mountain ranges due to radiating, configured, extremely low frequency sounds generated by valley specific winds.

The second concept is that of compass orientation. If there are environmental energy forms capable of providing a consistent heading to a potential long-distance mover, it should be able to evolve patterns of habitat transfer by which it only needs to head in a constant direction to make the next stage in the sequence. It might even be possible to change compass headings en route if the animal carries an internal clock, so that a migration course could be bent. Many stocks of salmon have to do this when passing through odd shaped lakes (Groot 1965). The experimental evidence indicates that much long-distance movement is by a compass sensitivity to such environmental energy forms as light, gravity, and magnetism. Light provides several such signals, including the configurations of stars, the position of the sun (clock needed), and the plane of polarization (polarizing filter needed). It has been shown that there are often redundant systems operable; when clouds obscure the sun, magnetic signals may be responded to.

One type of observation shows that another phenomenon is operating. Storm drifted bird flocks blown off route to England from the European coast, after a landfall and rest, can correct their compass heading and still make their usual winter quarters in Europe or Africa. They respond to the goal of where they are going. This is navigation in the sense used by humans for their long-distance movements over the oceans, where mariners are provided with a map locating harbors, but allowing the mariner to choose the route. The essence of navigation is that the migrants must have a position fixing system so that they can respond from where

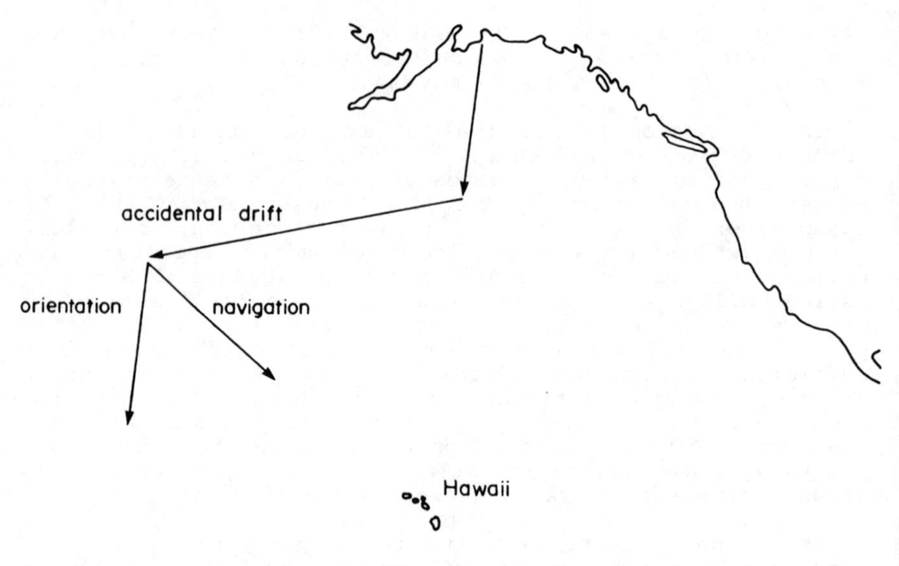

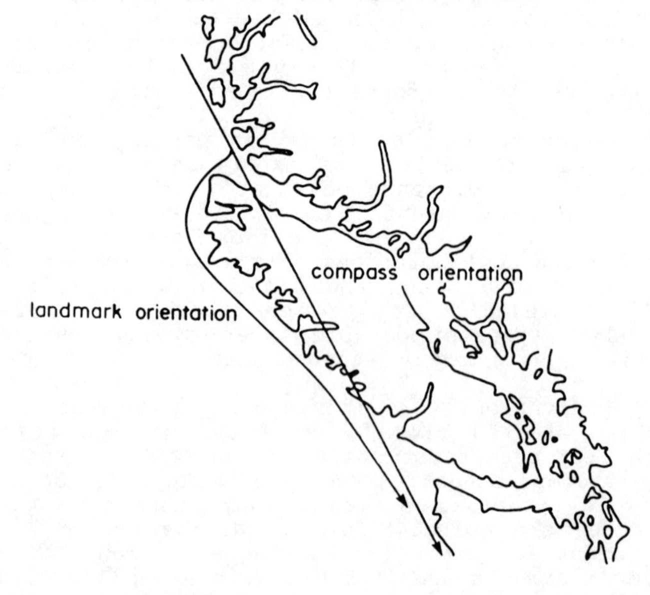

Figure 8.7. Orientation, navigation and pilotage.

they are by setting a heading to where they need to arrive. In very long-distance migrants this ability could be a response to star patterns, which appear differently at different points on the globe. The evidence for the interplay of navigation, compass orientation and pilotage is a fascinating story, and attracts much attention (e.g. Aidley 1981).

Homing

One of the problems in long-distance movements is when to stop. What are the animals responding to when they change behavior from the persistent movement of migration to the everyday activities of being in the right habitat? There are two obvious ways by which this can be achieved, but whether they encompass all the mechanisms is doubtful, even though this is a relatively little explored component of long-distance movements.

One mechanism is to rely on changed environmental stimulii, so that the releasers of the long-distance movement are no longer present, or no longer effective through being overwhelmed by more imperative stimulii. Thus a magnetotactic mud bacterium accidentally displaced to the water column and swimming down a magnetic line of force in the northern hemisphere (up in the southern hemisphere) will eventually return to the sign stimulii for mud instead of water, and can stop the magnetism taxis without risk (Blakemore and Frankel 1981).

The other mechanism is homing (Figure 8.8), the return to a place formerly occupied instead of other equally probable places (Gerking 1959). Equally probable means providing equivalent survival opportunities, except that the home occupier has the advantage of remembered occupancy. Homing provides a flexible termination orientation mechanism allowing a long-distance migrant to respond to an imprecise directional signal sending it in only the general direction of where it needs to be, followed by finely tuned orienting in the known right habitat. The latter comprises configurational landmark orientation, although there may be intermediate progressively finer tuned orientation sequences. Thus salmon, after making a coastal landfall solitarily from the high seas in the general area of the home estuary, appear to shoal up and head in the right direction along the coast. The sign stimulus is unknown. But at the right estuary home stream odor appears to become the sign stimulus and then all the fish has to do is swim up current against the home stream odor and it will bypass all wrong tributaries. (Lake transit is directed and so may involve other mechanisms.) At the home stream spawning bed the fish rarely overshoot, hence they must have switched off

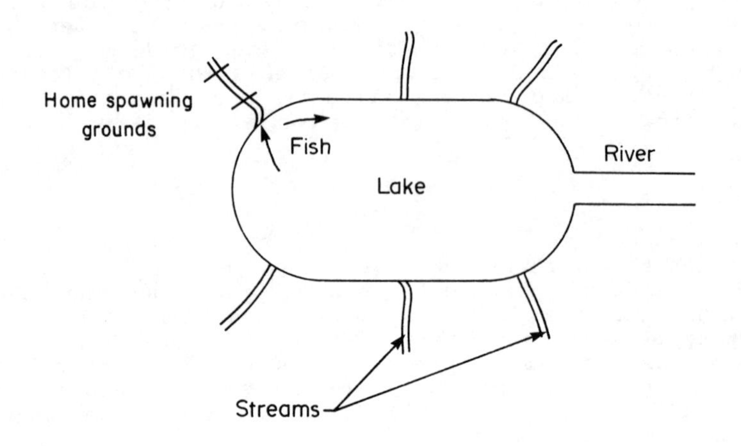

a. Trout

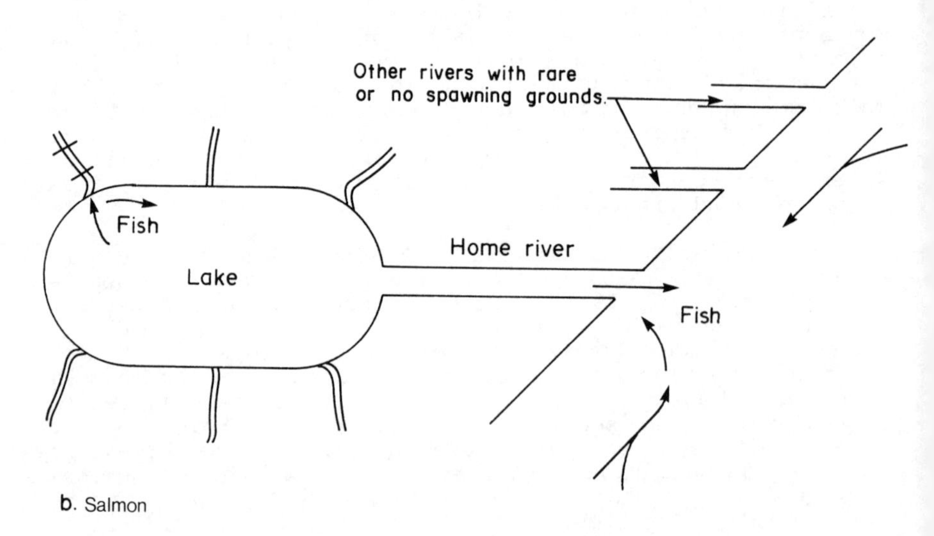

b. Salmon

Figure 8.8. The concept of homing as a pre-adaptation for long-distance movements.

migrating due to the effectiveness of the spawning bed stimulii prior to the disappearance of the odor.

Homing would so increase the probability of finding the isolated right habitat far upriver from the sea that it has to be an essential component in the evolution of anadromy in the Salmonidae from the freshwater territorial trout and chars. It is hard to believe that random river search would be an effective option, when the homing capability occurred in the primitive species. By equivalent extensions in other territorial or home ranging species it is possible to conceive shaping almost any extensive migration drawing upon sequenced, or mixed and redundant, environmental stimulii. Thus turtles can find and spawn on mid-Atlantic Ascension Island possibly through continuous homing on one or more of a chain of geologically temporary islands since continental drift separated the South American and African continents 100 million years ago.

Homing appears to be an essential adaptation to the very long-distance migrations embarked on by active, sensing animals with social organization involving territories or home ranges. It does not seem to operate in such stages as the drift larval phase for the transoceanic migrations of eels, and cannot be operable in species for which there is no capability of home recognition by individuals, i.e. no previous occupation, or no adequate sensors.

The obvious termination processes in long-distance movements direct our attention to the concept of the termination of any behavior. Is it by the disapearance of releasing stimulii, or the occurence of actual inhibitors? There are many unknowns.

Biological clocks and rhythms

It has long been known that some ants, returning from feeding, orient to their nests from the position of the sun. But the sun changes position, about 15° per hour, and some ant species cannot correct for this time dependent position shift of their beacon. They can be decoyed into the wrong direction by hiding them under a light-tight box for several hours, after which they keep the same angle to the sun, but it takes them the wrong way from their home nest. They do not have an internal clock. In contrast many birds, bees and even amphipod sand hoppers do have a biological clock. Some can correct their course by means of the sun's detectably changed position if delayed for more than a few minutes.

It was previously mentioned that fiddler crabs have a lunar clock which will constrain their emergence from burrows onto beaches to nocturnal low tides. The clock is maintained even when specimens are kept under laboratory conditions. All the periodic migrations shown in Figure 8.2

imply that the releasing stimulii might be internal, coming from the operations of an internal clock changing the physiology of the species and so priming it to respond to releasers for long-distance movement. Such clocks might be daily, lunar, or seasonal (spring, summer, fall or winter). They may even run for longer periods to initiate the four year cycle for sockeye salmon. Biological clocks appear to be widespread in the animal kingdom, capable of telling various cycles of time, and if inherent and accurate could answer many of the questions concerning how animals break into appropriate movements at the right time.

The evidence for clocks is compiled by keeping animals under constant conditions, the normal regular alterations of which might provide the signals for clock-based behaviors. Thus birds maintained in constant light or dark (or humans maintained in a free choice light-dark habitat) show a circadian rhythm (= about 24 hr) (Figure 8.9). There is an internal clock. However, the clock is not accurate. It is set too long and needs entraining by environmental stimulii. Left in the dark, or to free choice, the longer-than-24 hour cycle drifts backwards (or "free-runs") and the bird or human is eventually awake for its day while it is still night outside, and vice versa. Eventually the cycle is complete, and the experimental animals have lost a day. With some birds in the dark, even a daily flash of light about dawn will serve to entrain the clock.

The evidence for annual cycles is similar, but the key environmental property entraining the clock appears to be light-dark ratios, i.e. relative proportions of darkness and daylight. Temperature and other environmental properties signalling winter and summer appear to be too variable to be good entraining signals.

The environmental signals for other period cycles are much less well-known, and for some of the possible multi-year cycles there is little confirmation that they actually occur.

One type of evidence for annual periodicity is the nocturnal hyperactivity which occurs in caged migratory birds. This is called migration restlessness (Zugunruhe) and coincides with spring and fall migration periods. It has also been demonstrated in salmon. It is potentially a means of determining whether any abnormally timed movements are primed by a biological clock or seasonally changing environmental stimulii such as photoperiod, or both.

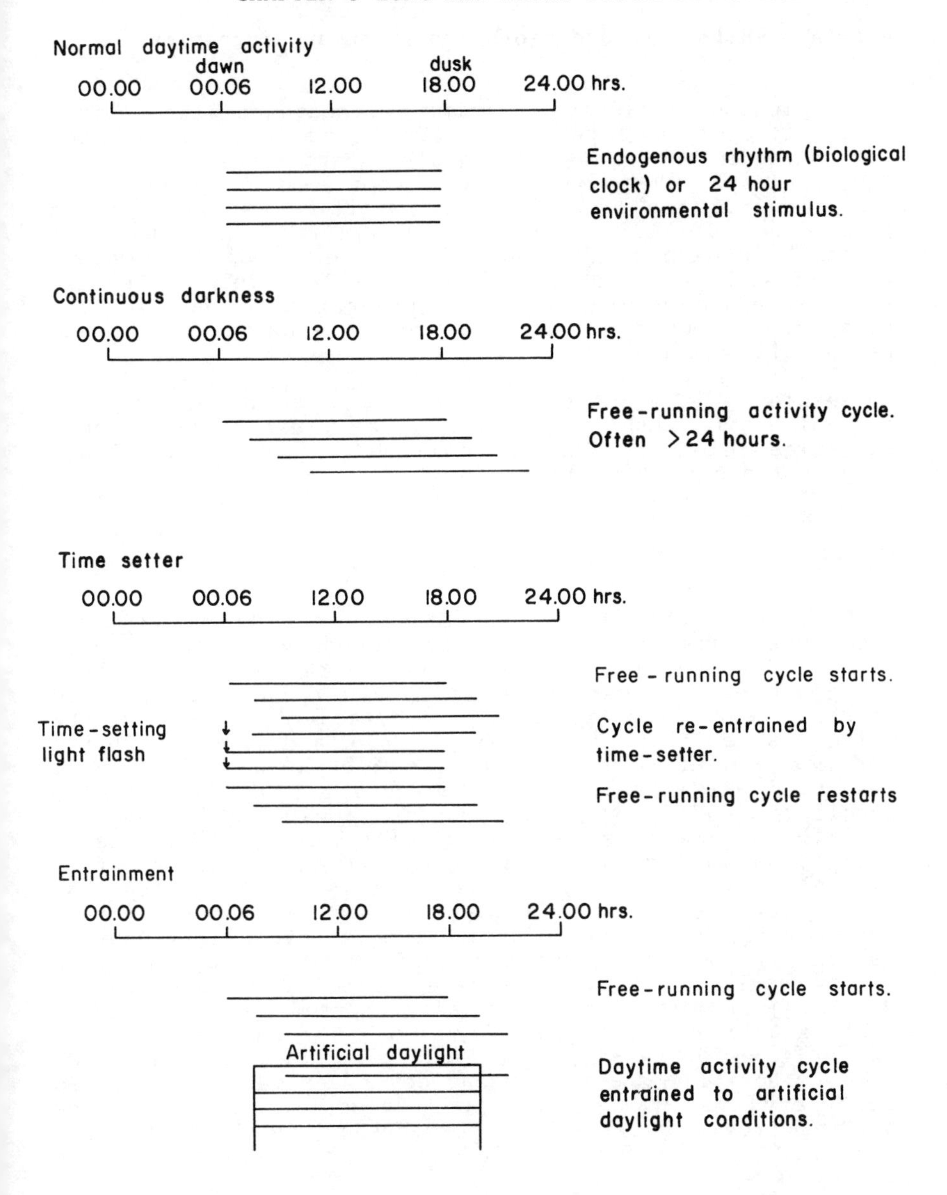

Figure 8.9. Biological clock shifts and entrainment. Normal daytime activity can shift backwards in continuous darkness, but can be reset by a single light flash or entrained by artificial daylight.

Jet lag, shift work and biorhythm (a human digression)

People show biological rhythms, although as with animals it is difficult to separate the effects of an internal clock from periodic environmental signals, such as dawn and night. Nevertheless, when people are kept under experimental conditions with a free choice of artificial light, as has been tried in light (and sound) proof rooms, and with subjects in deep caves, they put themselves on a stimulus-reduced version of their normal daily schedule. Usually their schedule becomes slightly longer than 24 hours. The human clock appears to drift as does the animal clock, and needs daily entrainment.

People also impose on themselves a number of physiological insults through the presence of their biological clocks. One of these is a consequence of our long-distance movements.

Jet lag

Jet lag arises from long-distance air travel (Figure 8.10) when passengers flying east-west or vice versa travel so far that they lengthen or shorten their days or nights. In flying east across continent or ocean the effect is to shorten the 24 hour day-night period, and in flying west the effect is to lengthen it. People have a daily body rhythm for metabolical processes (e.g. urine formation). As a result of air travel, the rhythm gets out of phase over a single 24 hour period. Since our daily body rhythm seems to approximate somewhere between 24 and 30 hours, we can travel east-west over a single time zone (about 500 miles) and feel virtually no effect. If we double this distance east-west (two time zones) or go farther, we may have to re-entrain our body rhythm to the delayed day-night cycle. It usually takes a day or so to do this and we are liable to feel sleepy early until we have re-entrained our body rhythm. Physiological disorders have been reported and many people feel vaguely unwell for a day or so. The rule of thumb is to allow one or two days for feeling normal again, although some authorities state that one day's recovery is required for every one or two time zones crossed. At least one airline provides a "peak performance planner" allowing 14 days for "full recovery" after a flight crossing 12 time zones.

Travelling west-east is different. The day-night cycle is shortened and the 24 hour rhythm must be brought forward. This is much harder physiologically, since the rhythm tends to drift the other way. However re-entrainment does eventually occur, even for west-east travel.

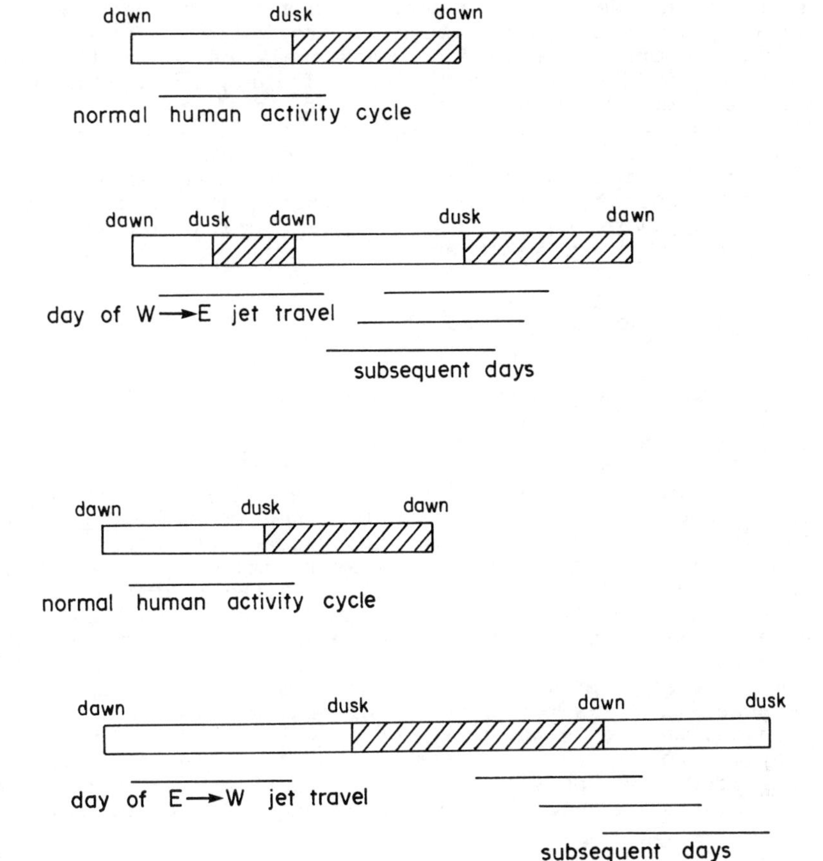

Figure 8.10. Jet lag and biological clock shifts. West-east jet plane travel exposes the traveller to a shortened day, and east-west travel to a longer day. Our daily activity period must be re-entrained, which can take several days of adjustment and physiological upset.

Airlines and travel conscious doctors now give advice to travellers, in part based on knowledge of the rhythm. It is basically that when you are about to travel east-west, for a few days beforehand go to bed progressively later, and when travelling west-east go to bed earlier. Try to get well started on the re-entrainment process for where you are going. There is also some folklore building up about not making important decisions immediately on arrival after a cross-oceanic or cross-continental flight, or trying to keep on your home time and ignoring the local day-night cycle. The latter only seems possible if the traveller could immerse himself in a light controlled habitat of hotels and meetings, and avoid seeing the outside environment. Other advice consists of sensible instructions such as removing stress by avoiding the complimentary drinks of air travel and frequent large meals and taking some exercise.

Sleep and other daily body rhythms

Jet lag affects the most obvious of human daily behavior cycles. Long-distance air travel puts our usual sleeping time out of phase with usual day-night environmental signals, including those which induce sleep. There is considerable controversy over why we sleep, but it appears to be necessary for us (except a few people) and animals (Meddis 1975). The main hypotheses are that sleep allows an inactivity period during which sense organs shut down, and the brain can collate newly acquired information and transfer it into its long-term storage facility. An alternative is that sleep allows physiological repairs to be undertaken. Finally, there is the posulate that for an animal sleep is a period of metabolic downtime adaptive because its needs for food are already met in a short period each day and it is advantageous for the animal to reduce its energy expenditure for the rest of the day. We have inherited the physiology from pre-human ancestors, and possibly extra functions have been added to the original.

We recognize sleep in ourselves, and in mammals and birds, with its behaviors of reducing sensory input, but there are also physiological characteristics: reducing metabolism to Basal Metabolic Rate (BMR), particular types of EEG waves, rapid eye movements (REM), and dreams. Some of these may occur in some other animals, but the common feature in what can be called sleep in many species is the drop in metabolism, and hence the conservation of energy.

There are other human daily body rhythms and considerable anecdotal information such as that some of us are morning people, whereas others are night people. This means that many people feel they are most active and alert, and work best, at particular times of day. These may be acquired patterns based on individual experience, but they may also change with age. They reflect the fact that the day-night cycle inevitably imposes physical changes on us and other animals, and that we can undoubtedly respond to it by the rhythmic expression of necessary activities. In animals it can be adaptive to do so, and sleep is one such adaptive behavior in terms of energy conservation.

Shift work

Shift work also interacts with body rhythms, especially sleep (Figure 8.11). The traditional shifts are those of seamen, who are professional long-distance migrants. They conventionally work four hour shifts every twelve hours. Watch keepers on the 8-12 shifts a.m. and p.m. are on a conventional wake-sleep rhythm. But the other two shifts, 12-4 and 4-8, require watch keepers to be attentive at times which for many people are sleep periods with attendant physiological lows.

The merit of mariners' watch keeping is that it is consistent. The watch keeper continues his pattern for as long as he is at sea, and as long as he holds his rank. Shift work on land does not have that opportunity for adjusting one's rhythm.

Industrial shift work which requires laborers to progress for example through day, evening and graveyard (approximately 12 midnight to 8 a.m.) shifts with intervening periods of a few hours or days, could hardly be worse attuned to human biological rhythms. Each time the shift changes the body rhythm starts adjusting but as it reaches adjustment a few days later, there is another shift demanded of it. When there are breaks of a few days in between, say four days after graveyard shift, the adjustment is to normal daylight during the long "weekend". If that shift is from graveyard to evening, just about the worst rhythm conditions possible are imposed on the worker. Anyone exposed to shift work should seriously consider how long they are prepared to tolerate the known physiological and psychological insults involved (Johnson et al. 1981; Mott et al 1965). Industries imposing such shift work, and unions concerned with workers' health, should consider this matter seriously. There is scope for much improvement in the timing of shift work in our industrialized society.

Biorhythm

Biorhythm is not biological rhythm. It is effectively a trademark for an industrial work system and its commercialization (Gittelsen 1975). Gittelsen maintains that when anyone is born, the trauma of delivery sets off three body rhythms of 23, 28 and 31 days. He calls them

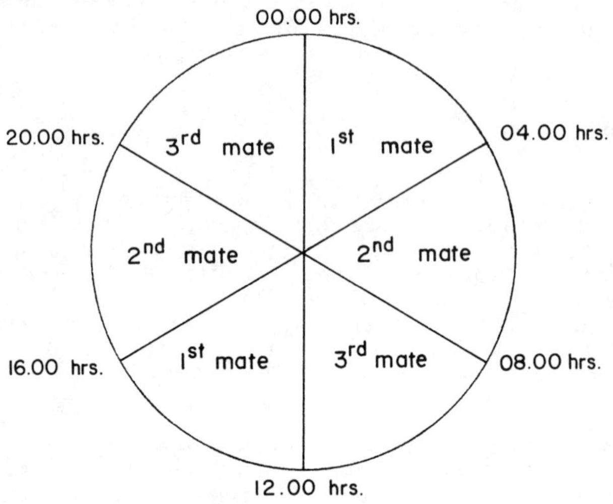

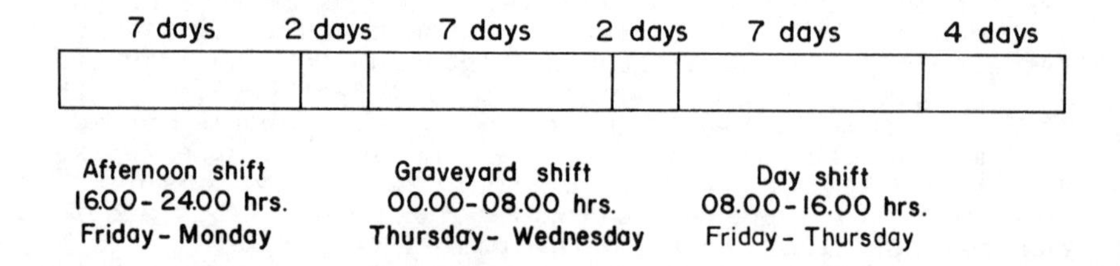

Figure 8.11. Types of shift work.

physical, emotional and intellectual cycles respectively. As the rhythms enter the downside phase of his curving graphs (Figure 8.12) they indicate bad times physically, emotionally or intellectually. If all of them are downside simultaneously, especially if they cross the graph's midpoint line together (critical days), then the individual is at the risk of disturbing one or the other of the three properties. He quotes terrible accidents which happened on

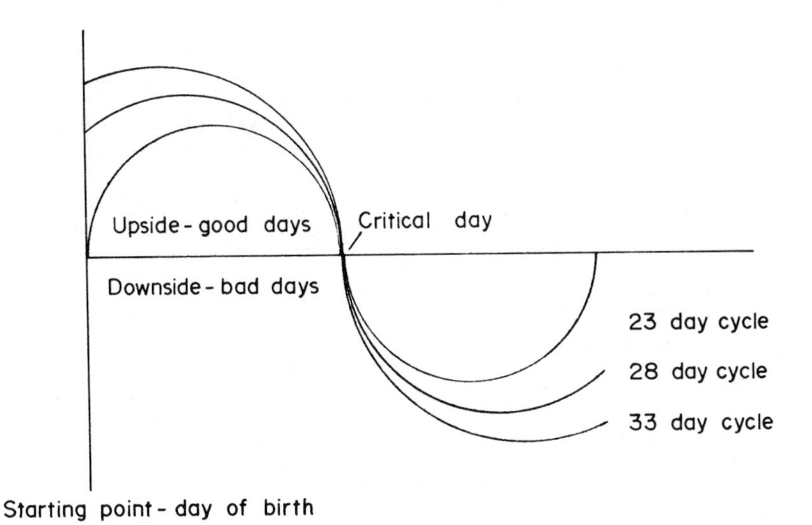

Figure 8.12. Downside, upside and critical days in biorhythm theory.

downside and critical days to famous people with known birth dates, and cites reduced accident rates in industries which have accepted his management consultancy theories. He also provides tables for his readers to calculate their up, down, and critical days. (They have to buy the book again every few years, since the tables in each edition are only good for about two more years.)

Biorhythm is a good example of non-science in behavior studies. There is no known corresponding physiological data on the rhythms (except 28 days in women). It is supported by documenting only the inevitable few positive coincidences. Unsupportive data is just not mentioned.

The best thing about biorhythm is that it provides behaviorists with an example of apparently authentic biological science applied to human behavior. Every behaviorist should read a (borrowed) copy of the book and devise a test to prove or disprove one or another of the implicit hypotheses, with the same rigor that they experimentally analyze other aspects of behavior.

Summary

Long-distance movements involve animals transferring themselves between habitats which are far apart relative to their size, strength, speed, and usual persistence in moving. As a result the animal reaches a new habitat, and new sign stimulii which were not perceptible at the start of the movement. Migrations are periodic long-distance movements between habitats, and the strategy of alternative habitats has been converged upon by a great variety of animals. Tactics for migrations may include active and drift movements, movements singly or in groups, and various period lengths from daily, to lunar, to annual and longer. There may be only one migration per lifetime or several generations may be needed to complete the round trip. Sequencing can draw upon landmark recognition (configurational stimulii), taxes (orientations), and navigation systems (point to point with awareness of intermediate locations). Directing stimulii range over many energy forms, but must be sequenced to accomplish the long movement. There is often redundancy and changing responses due to internal biological clocks.

CHAPTER 9. ENVIRONMENTAL IMPACT ON ANIMAL BEHAVIOR.

We impact on animal behavior in impersonal ways when we manipulate the environment to serve our own purposes. A simple but devastating impact of this kind occurred more than a century ago when Europeans first started building high lighthouses to guide ships on dangerous coasts. Many of these lighthouses were on the coastal routes of narrow front bird migrations, and during storms, particularly, streams of birds attracted to the light were hurled into the structure by the wind and killed in large numbers.

There are many other cases of mass or total destruction, ranging from African game animals to the dodo. Many of these are unexpected side effects of our actions because we are ignorant of animal behavior, and ecosystem checks and balances. But the impacts also arise because animals are unable to modify inflexible behaviors which had given us an easy strategy for exploiting them.

We need to understand the impacts of the past and the behavior of the animals involved if we are to mitigate the impacts in similar cases in the future. It is possible to gather information from many cases, and to show inductively that there are patterns (of behavior) which render some species particularly vulnerable to environmental impact from humans. With this understanding it is possible to show that certain types of impacts will happen to certain types of animals unless care is taken to mitigate those impacts.

Impacted behavior (case histories)

The case history concept is the environmental scientist's application of inductive reasoning. Case histories, properly collated and documented, provide examples of what happened. If there are enough case histories, patterns of impact emerge.

Hell's Gate

An early, little known, but dramatic case of catastrophic environmental impact occurred on the Pacific coast of Canada in 1913. It was two years since construction of the socially important intercontinental Canadian Northern Pacific Railroad had started in the Hell's Gate canyon of the Fraser River (Figure 9.1). Every year up to that time the river passed runs of Pacific salmon totalling in the tens of millions of fish. They were on their way to their home spawning beds from rich high seas feeding grounds, and were fished en route by commercial fishermen and native Indians alike. Every four years including 1913 a so-called "big year" upped the numbers to a hundred million or more salmon. The economic value to the small but growing western Canadian community was many million dollars annually.

At Hell's Gate from 1911, as the railroad bed was levelled, the blasted out rock was simply allowed to tumble down into the canyon and it eventually caused major slides into the river. Nobody thought anything about it. It was a quick, cheap way to dispose of the waste. Nature's way, no doubt!

Unfortunately, Hell's Gate was special. The river profile at that point was sufficiently steep, and consequent currents sufficiently fast and turbulent, that the millions of adult salmon could only swim past the jutting cliffs for about 100 meters on either side of the river, at a limited range of discharge levels. The increased load of rock raised the profile at many of these critical river discharge stages to just over the velocity and turbulence conditions which allowed the fish to pass. "All through August, September and early October the major portion of the run could be seen in the eddies below Hell's Gate, extending downstream for over ten miles" (Babcock 1914).

Another slide of materials occurred in 1914, and this overwhelmed attempts to correct the 1913 situation. Babcock and other fisheries officials managed to get some remedial action, but the amount of material to be taken out of the river, during a relatively short period of low discharge, was too much. From then on the spawning runs were too small to cope with increasingly heavy fishing, and the numbers declined until in 1945-46 fishways were tunnelled through the barricading cliffs so that what was left of the runs could pass.

The impact here was mediated through two behaviors: inflexible return to a home spawning ground and swimming behavior (type and speed limits). Pacific salmon have a steady swimming speed which approximates 1.5 body lengths per second (Ellis 1966a). Anything faster is alternated

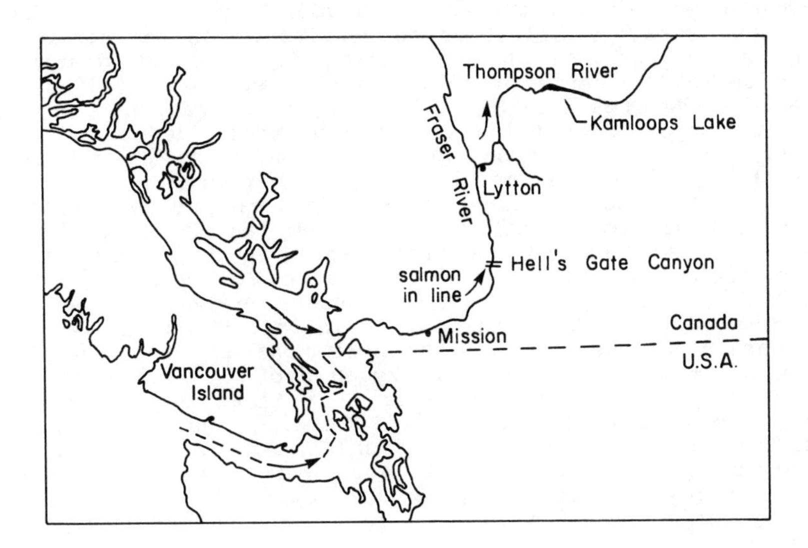

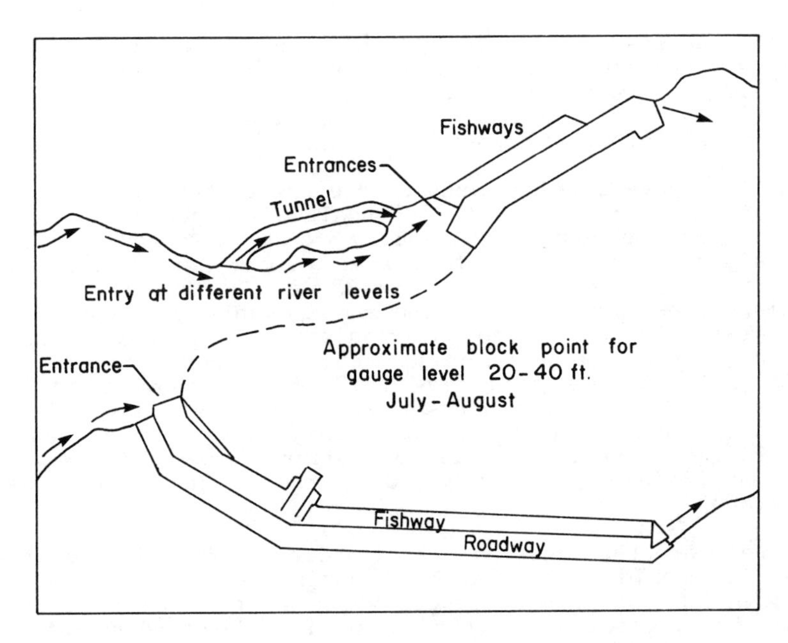

Figure 9.1. Location of Hell's Gate canyon on the Fraser River, and the point of impact on salmon runs.

with waiting periods, and if much faster puts them into oxygen debt like an athlete in a 100 meter sprint. Their swimming strategy is either to persist at an aerobic speed like a marathon runner, or sprint for a short distance. They sprint to catch prey or start a leap at a falls, but they use the steady swimming strategy on migration. In many rivers they can find low velocity water at depth and this allows them to progress upriver by swimming at 1.5 body lengths per second or less against currents at lower velocity (Appendix Figure A2.1). Thus they nicely balance aerobic migration with protection from land based predators such as bears and native fishermen. In deep fast rivers like the Fraser, they can only find low velocity water against which they can make steady headway at surface through eddies. They must often interrupt steady swimming to dart from one eddy to the next, and if necessary then rest briefly to come out of oxygen debt. To this day, in the Fraser and other big rivers, at times of large runs, salmon can be seen head-to-tail working their way up the narrow pathway of eddies available to them close to the river bank (Ellis 1966b). In such hydraulics they may not find energy assisting hydraulic jumps which assist and direct effective leaping at falls and rapids (Appendix Figure A2.2).

The Hell's Gate slides, even when modified by rock removal, reduced the already short period during which bankside migrants in their millions could jostle into position and then dart around the critical jutting rocks and rock piles into resting eddies beyond.

Salmon have suffered innumerable environmental impacts of this type. At Hell's Gate the scale of economic and social costs can be estimated fairly well. Thompson (1945) calculated that losses of sockeye salmon during the four-yearly "big" years meant losses of 279,000,000 in U.S. dollars from 1917 to 1942, and about double that for losses all years. Add losses of pink salmon, continuing losses of unrecoverable stocks after the fishways were built, subsistence values to native Indians, and convert to 1984 dollars. In economic terms alone, not counting non-dollar social values to natives and immigrants, losses can be considered close to the billion dollar level. This was one of the social costs of building the second trans-Canada railroad.

These are the kinds of social and economic balances which must be considered at times of environmental developments. Many of the older misbalances which we now try to correct were brought about by the lack of understanding of the biology, particularly the behavior, such as swimming patterns and speeds, of the resource species concerned. With salmon, the migrations are physically limited to narrow fronts in rivers, which are exaggerated by the selection of distinct pathways within the total water space apparently available to them. The costs of correction are very real, and are often externalized so that they fall on the public as a whole (the taxpayer) rather than on the developer and the user.

The passenger pigeon and the bison

Both the passenger pigeon and the bison were enormously abundant in North America when European colonists first spread into their habitats. Passenger pigeons could darken the sky with the extent of their flocks and the numbers of birds, into the hundreds of millions, packed together. Bison were so abundant that their herds could transform the plains into the appearance of a moving sea. No pigeons remain, and few bison.

The pigeon was blasted out of the air and their grouped nesting trees by arrayed shotguns, the firers of which had virtually no need to hide themselves from a species that had no programmed escape distance to humans. The minimal escape response and the formation of huge nesting colonies (850 sq. miles and 136 million birds in Wyoming in 1878) and narrow front migrating flocks were the two inflexible behavioral characters which brought the species into terminal contact with an expanding species, one with flexible strategies for predation and a need to "play", i.e. to kill for sport.

The near loss of the bison is more complicated. There was a progressive encroachment of intruding white men on its habitat, preying upon it. But the native Indian predator also adapted to introduced distance weapons of rifles, and to mobility through the horses of the immigrants. The consequent almost terminal decline of the species (25,000,000 to under 600) was halted by the environmental conscience and political concerns of later generations of taxpayers (Rensberger 1977).

The African elephant

Ian and Oria Douglas-Hamilton (1975), a husband-wife team of biologist-naturalists, have documented the way of life of the elephants in Tanzania's Lake Manyara National Park. The elephants have been progressively confined. Only some of them have learned the security of remaining within the park, i.e. most of the mothers (with their young) are not seen outside. The park is edged by farms which provide all sorts of accessible and delicious growing food. The farms are badly located so that they intrude over established elephant trails running between one end of the park and the other.

In addition, male elephants characteristically wander more than females, and there are established trails for males leading out of the park into nearby high plateaus which have been taken over for farming since the park was legislated into being.

Conflict between farmers and elephants is now inevitable because the good intentions of government conservationists were not based on adequate knowledge of the matriarchal social organization of the species, the wandering of males in small groups, and also the periodic movements of the mothers for a change of food and other inexplicable reasons.

In addition confinement in a small area, isolated from an informed concerned protective public, renders the elephants susceptible to poaching. Ivory is valuable, poverty is rife, and the temptation great. The park is now overgrazed, and some elephants must be eliminated by transfer or culling. But conservationists are loathe to cull, and transfering elephants is difficult. The elephants of Lake Manyara are a good example demonstrating that conservation and enhancement measures can also have their bad impacts if biology, including behavior, is not known well.

The hunting of the whale

The large whales of the world were hunted almost to extinction during the 19th and 20th centuries. It was not easy for men to do this. They had to work at it. Even large whales in schools would be difficult to locate on the enormous plains of the high seas if there were no patterns of behavior to guide cunning naturalists in the form of whaling skippers.

We now know that many whale stocks follow distinct migration routes between seasonal habitats (Lockyer and Brown 1981). Thus the Hawaiian-Alaskan humpback whale spends winters in the channels between the Hawaiian islands, and summers on the coast and in the productive fjords of Alaska. This must have been a difficult pattern for whalers to deduce, and thus probably accounts for this stock remaining.

It now appears that the large baleen whales had extensive north-south regular migrations which could be intercepted by whaling boats. It has even been postulated that some species simply rode the circum-antarctic current around the pole, taking several years for a global circuit, although this now seems unlikely.

The important component of this hypothesis (which is now very difficult to test) is that whales did not roam randomly through the high seas. There are environmental patterns to food production in the oceans, and the whales evolved systematic responses to these, thus increasing their survival and growth chances. Unfortunately for them, a new predator arrived on the high seas, some of whose individuals had the ability to decode information in the ecosystem about whale migration patterns. They introduced a ship-based ambush strategy to which the whales had no avoidance pre-adaptation such as an escape distance beyond harpoon range. Note that only a few humans needed the intelligence to decode the migration information for substantial numbers of social followers to adopt the same pattern of predation.

The impact of humans here is mediated through the inflexibility of the whales in critical components of their behavior, i.e. predictable migration routes, and short or no escape distance to men in ships.

The dodo and the moa

The dodo and the moa were large flightless birds which disappeared soon after human contact was made with them. Their contact was with historic and prehistoric immigrants to Mauritius and New Zealand respectively. Little is known about the birds, but since they were flightless and large, we can deduce that some of their behaviors followed the patterns of other large animals. A behavior common to large animals inhabiting regions without large fierce predators is a short escape distance, if any. When nothing can attack you, a personal escape distance can disappear from your repertoire. If there are other large but harmless animals around, you may retain an escape distance simply so that you avoid the other. Physically it may bump you around a bit even if it does not attack you, or it may compete for available food. So the dodo and the moa probably had only short escape distances, if any, to other large animals, and none at all to small introduced egg eating rodents. When humans appeared on the scene, the situation was very different. The immigrants were armed with distance weapons (spears at least), the intelligence to coordinate group hunts, and carried pests (rats and dogs) with them. We can postulate that the dodo and the moa had no counter strategy for the arrival and distance attacks of humans over the few generations before they became extinct, nor against egg eaters. As with whales, the lack of a counter strategy to survive the impact of humans, and insufficient breeding time for the evolution of a strategy, started population losses which led to extinction.

Environmental impact assessment

The scientist's response to the challenge of environmental impact has been to develop a set of procedures for assessing and predicting impact at new developments so that the worst effects can be reduced to a level which society will accept as a reasonable balance for whatever benefits the development will provide. For example, at Hell's Gate it might have been a reasonable balance if only in two years out of four the earliest and latest runs of salmon were greatly reduced in return for the benefits of a second transcontinental railroad linking the west coast to eastern Canada. This leaves aside the option of providing compensation for the affected fishermen. The latter is a social issue, not one of animal behavior.

Some of the early assessments have been very useful in indicating how others should be made, and hence impacts reduced. Others have not been so successful; developments have gone ahead and impacts have been worse than predicted.

The Porcupine caribou herd

The Porcupine is one of the great caribou herds in Canada, approximating 100,000 or more animals. It occupies the Yukon Territory and Alaska by migrating between a summer range in the coastal arctic plain and a diffuse winter range to the south (Figure 9.2). It has survived the hunting pressures of Indians and Inuit provided with rifles due to its isolation from the native population centers largely to the east along the Mackenzie River and delta, and to the west in Alaska. Now the herd is being stressed due to the availability of motorized sledges, but even more due to industrial developments. Of these, the great oil rush of the last two decades, following the more westerly Alaska find and pipeline, is the most serious. In the late 1960s and early 1970s there was an industrial proposal to pipe oil and gas to southern Canada and the U.S.A. via pipelines along the Canadian arctic coast and the Mackenzie Valley. These would have had such dramatic socioeconomic impact that Mr. Justice Berger was authorized to conduct an inquiry into the proposals. The Porcupine herd ranged within the area of interest of the Berger commission, and so a train of assessment investigations was set into operation. Industry, government, groups of concerned citizens and academics, the last working independently or retained by one or another of the involved parties, all contributed, often with different opinions among themselves on the accuracy of the information, impact predictions based on it, and interpretation in terms of mitigation to be followed by the industry if the pipeline were to be built. As a result of

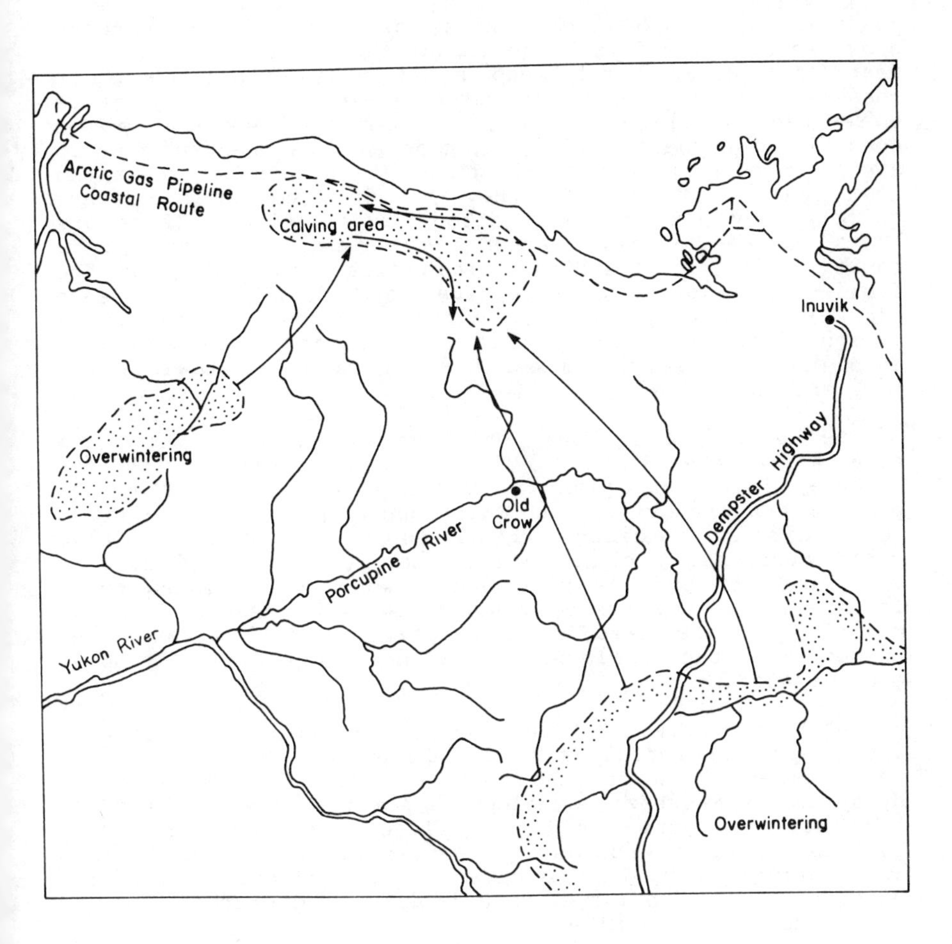

Figure 9.2. Range and migrations of the Porcupine caribou herd.

this controversy there developed a reasonable understanding of the Porcupine herd, its ecology and behavior, and especially its migrations and sensitivity to impact (Berger 1977).

In the spring, the diffusely spread caribou females and juveniles start moving north from their winter range, feeding as they go. The migration speeds up and the herd aggregates near the north coast, concentrating in the coastal plain. Here the calves are born during a short slowdown of a few days. The infants are precocial, on their feet soon after birth, and must remain close to their mothers to imprint on them, and vice versa. The herd starts a reverse migration eastward and remains packed into the coastal area where, if the concentrations are subject to disturbances by predators (or humans), infants get separated from their mothers. If they do not find their mothers quickly they die from starvation or predation. There are bears, wolves, foxes, marten, wolverines, jaegers, owls, hawks and falcons waiting. The males arrive to find the females more or less receptive, and during the reverse migration rutting starts. The movements may become irregular at this time, but in late summer the migration is more directed as the mixed herds move south towards the treeline and spread out.

The Berger commission identified the sensitive time and area where the Porcupine herd would assemble with its recently imprinted infants. The area was on the track of pipeline construction, and maintenance would require crews to be carried by disturbing helicopters. There were also other concerns, such as whether caribou could physically cross over or under a pipeline or would use bridges built for them. The key issue, however, was the assembly of the majority of the herd in a relatively small area during summer, when the development would be most intensively built and operated.

The Berger assessment was that there would be substantial impact on the Porcupine herd, which with other predicted major impacts (on resident natives) led him to recommend delaying the Mackenzie development for ten years and finding a substitute route for the pipeline, i.e. not across the caribou calving grounds. The assessment produced much information, and an accepted recommendation that the existing social balance should not be disturbed by the development at that time.

Resource management cases

Many biological resources are managed; they are managed in the sense that there are attempts to maximize the yield or crop of fish, wildfowl or game and allocate it to particular harvesters. Management can be done well in that the yield may be maintained at a high level and spread in socially acceptable ways between commercial, subsistence and recreational users. But management can also be done badly. There can be not only poor yields, taken largely by privileged groups, but unpredicted side effects may be generated.

Salmon provide many examples of both good and bad management, arising from understanding or misunderstanding the behavior of the fish.

The design of fishways to pass salmon around natural barriers blocking spawning migration or man-made river alterations blocking or misleading the fish, has been modified over the years. Originally, fishways consisted of ramps discharging water which was often too fast, too turbulent and too shallow for fish to swim up, or of pool and weir steps with badly located entries and a hydraulic jump which did not aid the migrants in their leaping. Clay (1961) summarized improvements in fishways reached pragmatically by engineers, and only subsequently did biologists demonstrate much of the behavioral reasons for their success (Stuart 1964). Salmon choose routes through water space which has the maximum stimulating effect so that fishway entrances need to be right at the blocking point. They move by steady aerobic swimming interspersed with oxygen debt darts, and need fishway eddies for resting. They can derive an assist in leaping from a well formed hydraulic jump. They have a maximum dart speed which approximates 10 body lengths/sec. Water velocity coming down a slot fishway must not exceed the maximum dart speed of the types of salmon ascending. The good resource management here was the imaginative consideration by engineers and administrators for an intuitive perception of the abilities of the fish. Biologists provided the backup information later.

Through a second development, that of spawning bed enhancement, salmon management has fallen into problems of unpredicted side effects. There has been a recent program in Canada to improve spawning beds by controlling gravel, water flow, operational tidiness (i.e. eliminate washouts, fallen trees, beaver dams, etc.), and number of spawners. So far so good. Such enhanced beds improve the output of young salmon enormously and the fish go off to sea. The unpredicted side effect appears on their return at fishery time. Salmon return to their home rivers, and prior to river entry each river's stock schools up for mass entry. Where several stocks school together and one or more has been enhanced and allows heavy fishing, other unenhanced stocks, particularly if they are small, can be totally or virtually fished out. So small runs near enhanced stocks suffer.

In short, developments intended to be beneficial are likely to have bad ecosystem side effects unless good impact assessment precedes the development.

Sensitivity matrices

With large numbers of impact assessments now being made and reported, it is possible to reason inductively about the accumulated information to show circumstances under which species are sensitive to impact. These sensitivity analyses then can be used deductively at specific sites to predict areas or times of impacts.

Such a sensitivity matrix can be developed for western North American ungulates (Table 9.1). A matrix of impacting developments and habitats occupied by particular ungulates is constructed. By simply summing the number of impacts for each habitat and development, indicators are provided for the worst case situations. Thus roads and highways affect many species in various ways in many habitats. The effects may be disruption of migrations due to traffic, actual kills of animals, increased access for hunters, and the provision of ambush habitat for predators (wolves are particularly good at using roads for ambush through otherwise difficult terrain). In terms of habitat sensitivity mature forest in valley bottom is particularly likely to be affected by almost any kind of development there. The species dependent on that habitat will also be affected by loss of optimum conditions or reduction to patches with high risk travel between them.

Sensitivity analysis can also be applied to different stages in a species' life cycle. Table 9.2 provides a matrix of salmon stages which are passed in river estuaries and a series of industrial developments which are likely to occur in such estuaries. In this case, kinds of processes within the impacting situations are identified by the matrix. Thus the leaching of toxins from wood products or wastes is a component which needs addressing at many developments, at sites where juvenile salmon concentrate for passage or feeding, or adult salmon concentrate for spawning.

Impact sensitivity matrices can be developed in many ways to direct attention to species, habitats, and life cycle stages. They could be used to direct attention to other biological properties such as migration route selection, breeding concentrations, time of year, or behaviors which appear to render animals particularly sensitive to impact from human developments. The technique is available, and should be used widely for guidance in Environmental Impact Assessment (EIA).

Table 9.1. Western Canadian habitat types, ungulate species commonly occupying them and types of impacts which affect each habitat (developed by B. Dixon).

Habitat	Species	Impacts										Total
		Hydro dams	Settlements	Logging/fire	Mines (all)	Agriculture	Roads/highways	Pipelines (all)	All terrain vehicles	Aircraft	Hikers	
Mature forest, valley bottom	moose, elk deer, sheep	X	X	X	X	X	X	X	X		X	9
Mature forest, not valley bottom				X	X		X	X				4
Open mixed forest	deer, elk			X	X	X	X	X	X		X	7
Early succession forest	moose, elk, deer					X	X		X			3
Grasslands, prairie	deer, bison, pronghorn		X			X	X		X	X	X	6
Open meadow, rangeland	deer, sheep, bison, elk				X	X	X			X		4
Arid and semi-arid shrubland	deer, sheep				X	X	X				X	4
Marshy shrubland	moose	X			X	X						3
Alpine	goat, sheep, caribou				X		X	X	X	X	X	6
Subalpine	sheep			X	X		X	X		X	X	6
Tundra-interior	muskox, caribou						X	X	X	X		4
Tundra-coastal	muskox, caribou	X	X				X	X	X	X		6
	Total	3	3	4	8	7	11	7	7	6	6	

Table 9.2. Example environmental impact matrix for salmon in an estuary (developed by C. McKean).

Salmon Resource Impacts	Development Actions							
	Lumber mill	Log storage in booms	Intertidal dredged channel	Freight wharf, rail line	Log haul out	Waste storage	Wood pelletizing plant	Pulp mill
Downriver migration by juveniles - diversions and toxicity	Fill, Leachate	Leachate	May assist migration	Fill, Leachate	-	Leachate	Fill	Fill, Water extraction, Waste discharge
Feeding habitat for juveniles - destruction	Fill, Leachate	Scouring, Wood wastes	Increased depth	Fill, Leachate	Scouring, Wood waste	Leachate	Fill, Leachate	Waste discharge, Water extraction
Upriver migration by adults - blockages	-	At certain tides	May assist migration	-	-	-	-	Water discharge, Water extraction
Spawning ground - destruction	Fill, Leachate	-	-	Fill, Leachate	-	Fill, Leachate	Fill, Leachate	Fill, Leachate

Working principles for assessment of environmental impact on animal behavior

The generalized process for EIA can be modelled (Figure

The Development	The EIA
1. Conceptual design	1. Baseline surveys
2. Obtaining rights - purchases, permits	2. Pre-development monitoring
3. Environmentally mitigated design	
4. Construction	3. Construction monitoring
5. Commissioning	4. Commissioning monitoring
6. Operations	5. Operational short-term monitoring
	6. Operational long-term monitoring
7. Closure (temporary)	7. Closure monitoring
8. Closure (permanent) (Reclamation)	8. Recovery monitoring (Reclamation monitoring)

Figure 9.3. Flow diagram of the process of EIA.

9.3). It shows that a total assessment can be divided into

a series of steps, only some of which are commonly understood as EIA. Other stages may be considered "baseline monitoring surveys" and "hindsight reviews". The terms can be used in several different ways. For our purposes Environmental Impact Assessment (EIA) is the higher level term, encompassing the others. Baseline surveys are the preliminaries which establish the state of the ecosystem prior to a development. Monitoring is periodic surveying to determine changes, and to feed back information so that the ecosystem can be quality controlled by fine tuning operations of the development. Hindsight reviews are periodic appraisals of the situation. In some cases, regrettably, hindsight reviews are introduced *de novo*, without prior assessment of any kind. This used to occur frequently during the 1960s and 1970s at sites of established impact, but hopefully will recur less and less as older sites are brought as close as possible to standards of impact assessment and control at new developments.

The process of EIA

EIA should begin at the time a new project is put into the concept design stage. The corresponding baseline surveys document the state of the ecosystem, particularly any resources which are of value to the local community or broader regions, such as fisheries, wildlife, rare species, historic and beautiful sites, and other competing industrial use. Information is fed back to the design engineers so that the design concept can be modified to minimize obvious impact on major resources, e.g. minimize toxic smoke plumes destroying forest canopy habitat and dependent insects and birds, and the discharge of toxins to trout streams. In short, assessors have the opportunity to introduce design constraints before the blueprinting of designs and before construction. After concrete is poured, design changes are expensive.

In the meantime, rights for the development have to be obtained by purchase or permit. This is the time to convert baseline assessments to pre-operational monitoring. This stage provides a measure of pre-existing variability in relevant parts of the ecosystem, so that subsequent impact changes are not confused with changes which would have occurred anyway. Caribou migrations may follow different routes from year to year, and a change immediately after a pipeline is built may have nothing to do with the pipeline. Baseline surveys and pre-operational monitoring together provide the data for before-and-after comparison.

The blueprinting stage providing instructions to contractors virtually prevents major design changes, and from then on only fine tuning of operations to mitigate impact is possible without major, and hence resisted, expense. Pre-operational monitoring must be done within a set term, or the opportunity is lost.

At the time of construction, special monitoring is needed to quality control the construction process. Typically, space is cleared for construction and access, and road building especially can impact stream and even river habitat.

After construction, operations should be tested during and immediately after commissioning. At this stage unexpected short-term effects such as surfacing sewage plumes attracting sea gulls may appear, or there may be deviations from permitted environmental constraints, such as with toxic or visible emissions from smokestacks. Commissioning requires special intensive monitoring of possible short-term effects, so that operational performance can be improved. The short-term monitoring should be continued.

Once routine operations start, monitoring should change to data gathering on possible long-term impacts. Do some species gradually die out, or move away? It is to be expected that long-term monitoring tests and sampling designs may be substantially different from those monitoring short-term effects. As operations continue so there may be a series of hindsight reviews which assess what the impact has been, preferably contrasting this with objectives or permit conditions stated prior to operations (Mitchell 1977).

It should not be forgotten that industrial plants have a lifetime, like animals, and will eventually cease operation. Progressively more and more industries, on closure, are being required to restore their sites to a productive ecosystem. This reclamation also needs monitoring and quality controlling, during temporary closures and the final shutdown, so that recovery is adequate and efficient. Mine sites have traditionally been required to reclaim stripped land for game or farm habitat, and this requirement is now being placed on other industries.

At all these stages, investigation has traditionally been ecological in that it has determined the structure of the ecosystem (physical, chemical, biological, habitat) or the rates of processes such as primary production and phosphate biogeochemical cycling. However, certain impacts are mediated through the behaviors of the ecosystem species. Animals can move away, or refuse to do so. They may even aggregate in unusually desirable (apparently) conditions in which toxins insidiously affect them. Some guidance for investigating such behaviors is needed. This follows.

The working principles

Principle 1. *Determine if stationary mass assemblies occur.* Where a development is proposed, any assembly of an animal species places more individuals at risk than if the species is dispersed. Stationary assemblies on land are easy to identify, as was the Porcupine caribou herd in its summer coastal quarters. But such stationary assemblies may occur underwater, in lakes or coastal area, and need detection by remote sensors (Harden Jones 1968). Care must be taken to examine the area at the right time of year, which may be during the mating season, at birth, hibernation, or juvenile care season. A stationary assembly may occur for a limited time only, and the investigators must be there at the right time.

Principle 2. *Determine if mobile mass assemblies occur.* Narrow front migrations of such animals as ducks and geese create moving mass assemblies. Conversely the assembly may stay put but the individuals are continually changing as they squeeze through a route bottleneck. Mobile assemblies are far more difficult to detect than stationary assemblies since either they change position through space and time, or few animals may be present at any one time at the critical bottleneck. This may be in spite of thousands or even millions of individuals passing during relatively short periods of a few days or weeks. Care must be taken to observe in the right places, at the right times, and especially for a long enough period; and to follow individuals passing in order to get accurate counts.

Principle 3. *Determine if the species shows any inflexible behavior.* Species having inflexible behavior can be at risk. Species like the sage grouse with its traditional mating leks do not readily adjust to intruding farms and highways by moving elsewhere. Inflexibility can occur at many different stages in the life cycle, not just during the courtship requirements of breeding. Inflexibility in homing behavior and flight distance to man renders animals particularly susceptible to impact.

Principle 4. *Know life cycles and ethograms.* The three problem types of behavior listed above can occur at any stage in an animal's life cycle, and for a range of activities including feeding and breeding. The impact assessor should ensure that he or she has a knowledge of the life cycle of important resource species; their feed stock, predators and competitors; and a reasonably detailed ethogram for each species at each life cycle stage. An incompletely known life cycle may indicate either a narrow front, impactable migration, or a hidden stationary mass assembly within the area to be impacted.

Principle 5. *Know ecosystem networks.* An ecosystem network analysis should be conducted or at least postulated for important resource species so as to show interactions with predators, prey and competitors. With such a network analysis, side effects due to populations coming into new balances may be predictable, or at least can be postulated and monitored. Monitoring provides opportunities to test reasonable hypotheses of impact.

Principle 6. *Collate data on other ecosystem stresses.* Often an industrial development proceeds in an area which is already subject to stresses from prior developments. The impacts from prior developments need to be noted in analyses 1-5 for the present situation, and original conditions must also be postulated or documented. The changes indicated can be guides to further changes which will result from the new development.

Principle 7. *Mitigate by avoidance, not continuing action.* Impact should be mitigated as far as possible by avoidance through the design of the operation. The alternative of continuing action, such as detoxifying waste streams, and building and maintaining landfalls, is less satisfactory since the action will have continuing operating costs and problems of quality control. Sooner or later something will go wrong, and then there may be a catastrophic impact rather than a chronic one to which the impacted species may have partially adapted. Often corrective responses such as oil spill cleanups will themselves go wrong or be delayed.

Summary

The human species impacts on animal behavior impersonally through developments affecting the environment as well as through personal interactions (Chapter 6). Many cases of environmental impact through animal behavior are known, from coastal lighthouses which lethally attract migrating bird flocks to farming encroachment which reduces game. The ecosystem side effects of developments can be reduced by the methods of EIA, data gathering and feedback to regulatory authorities or design engineers. Impacts from developments are most likely to occur where species form mass assemblies, either stationary such as those on breeding grounds or mobile in the form of narrow front migrations. Inflexible behavior also renders impacts likely because of the inability of a species to adapt to human encroachment. Impact assessors at specific sites should have knowledge of the important resource and ecosystem species, and their possible mass assembly and inflexible behaviors. It is desirable for an assessor to complete life history diagrams and an ethogram for these species to indicate missing behavior which might allow better impact prediction. Mitigation is best achieved by avoidance of impact rather than continuing action since the latter needs quality controlling. Accidental breakdowns can cause catastrophic impacts instead of chronic situations to which a species is partially adaptable.

CHAPTER 10. EVOLUTIONARY (ULTIMATE) CAUSES.

For the biologist, the ultimate cause why living things are the way they are is evolution. Some forms of inherited characters (e.g. a faster response than normal to predators might be genetically determined) have been more effective than others. The characters have been better adapted to the demands of the ecosystem. They have had greater survival value. There are two processes which bring this about. One is change in the inherited material; the other is natural selection.

The concepts of animal behavior and evolution currently have a circular relationship to one another. Animal behavior is both explained by evolution (the ultimate cause) and is used to help explain evolution -- by selection. There is real danger of circular argument, and it is important to anchor one's understanding of causative theory in one's own or others' observations.

Change in the inherited material may be a biochemical change in the DNA make-up of the genes (mutation), or the result of a rearrangement of the biochemical order during meiosis and sexual reproduction. The genetic change alone though is not enough. A new character may be produced, but it has to prove its worth in competition with other similar characters held by conspecifics or other species if it is to remain in the gene pool. It also must prove its worth by compatibility with all the other more or less well adapted characters held by its bearer. A faster dart escape but quicker immobilizing fatique might not be an advantage against some persistent predators.

The second ultimately causative process, natural selection, does the proving of new characters through the effect of competition. There is a perpetual struggle for existence, avoiding being eaten or being destroyed by environmental stresses, and finding food or mates. Some animals are more likely to survive and breed than others. The balance of their characters is better adapted. They are fitter.

These two processes are in action together. Mutation and reordering genes produces the inheritable variety of characters. The processes are random. Mutation does not produce a directed series of changes. The directedness as we see it in hindsight comes from natural selection. If a constant ecosystem pressure, such as the environmental temperature change in Figure 1.6, continues over enough generations of a species, the individuals best able to survive and breed in the new balance of optimum conditions will displace individuals without that balance of adapted inheritable characters.

Any inherited change in a behavior is filtered through a screen of natural selection not on its own, but through the way it integrates with the animal's total package of behavior. In this total context we can expect that many inherited changes will be maladaptive, i.e. degrading existing Evolutionary Stable Strategies (ESSs - Chapter 1), and that new ESSs may well involve fairly substantial but rarely achieved behavioral changes.

The evidence for evolution, and its two processes, was initially produced inductively by the enormous collations of Darwin (1859, 1964) and his contemporary Wallace, who extended ideas of their predecessors. Darwin formulated the concept of natural selection, but it was left to others to establish the details of random mutational change (see any introductory genetics text). The evidence is largely structural and chemical, since behavioral characters are less easily described: behavior is variable in many subtle unmeasurable ways, and is affected by multiple genes and individual experiences. Nevertheless, the behavioral evidence must be considered by the biologist. There are social implications inherent in the concept of behavioral evolution. Some people are resistant to believing that behavior can evolve, or at least that behaviors of humans can be derived, however remotely or partially, from the survival ability of their animal or pre-historic human ancestors.

There is reason for such skepticism. There are powerful people who claim that survival of the fittest is an appropriate way to organize human social and political affairs. During the oil gluts, shortages and price spirals of the last decade, the president of one of the larger oil companies quoted Darwin to justify ruthless economic competition. This is social Darwinism. It has been used to justify oppression by powerful people since Darwin's time and when it appears it needs to be seen for what it can be - a rationale for human exploitation.

Conversely, there are others who are so fearful of social Darwinist implications and biological determinism (that human behavior and social organization contain inherited components) that they strongly resist investigating the topic. Behaviorists must realize that they are working on socially controversial matters when they apply ideas concerning ultimate causation to humans, and they should understand the background to the controversies. A discussion is given by Ruse (1979).

For many people one or other form of "creationism" is part of the ultimate cause why living things are the way they are. Although some forms of "creationism" may be compatible with natural selection (and allow species change after creation) directed genetic changes implicit in "created" species would seem to be inherently incompatible with random mutational changes and chromosome reordering, and scientifically untestable.

There is considerable, and controversial, theory about the processes of evolution in general, and the lines of descent for particular monophyletic taxa. Both processes involve the separation of sibling species, but the problem of apparent quantum jumps in anatomy between substantially different taxa, while maintaining a behaviorally viable animal, must also be considered. The generalized theoretical concepts, such as inclusive fitness, must be applied to phylogenies at species and higher taxa levels, otherwise they are simply just academic speculation. This is particularly so in a context of geological time and prehistoric ecosystems; fossils need understanding as "missing links" and as viable animals in their own right. A special role of the behaviorist in this context is to consider hypothetical ancestors and fossil forms as viable, behaving animals.

Evidence for the evolution of behavior

Homologous behaviors

Lorenz (1951) provided a detailed model of the concept of homology applied to behavioral characters (Figure 10.1). Homology generally refers to anatomy. It means that some structures in different animal taxa have a common evolutionary origin. The paired fins of Teleost fish and the legs of reptiles and mammals are homologous. They are descended from the two sets of paired appendages of the common vertebrate ancestor. The wings of birds, the forelegs of mammals, and the arms of humans are homologous. They are descended from the forward pair of the vertebrate appendages. The science of taxonomy, which classifies animals in phyla and other taxa, is based on the recognition of homologous structures and their separation from analogous (convergent) structures, i.e. similar looking since they are adapted to a common function. Thus the wings of insects are analogous to the wings of birds; they are not homologous.

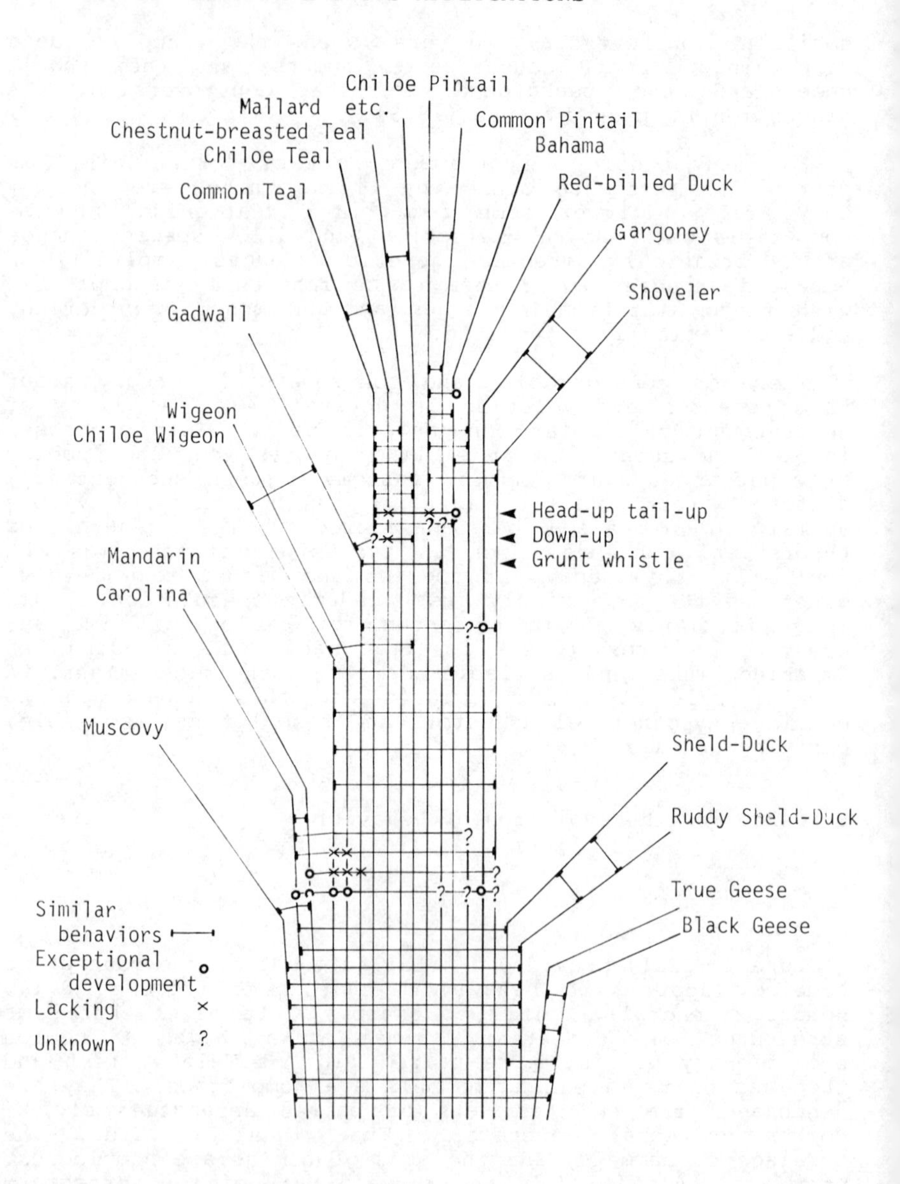

Figure 10.1. Lorenz's shaving brush model of Anatid behavioral similarities. Three mallard courtship behaviors shared with other species are indicated. For other behaviors see Lorenz (1951).

As an introduction to the behavioral applications of these two concepts, we can say that both birds and insects have converged on flight as a locomotory behavior, and have evolved analogous flight behavior strategies. Various birds and insects deploy flight in analogous ways to suck nectar from flowers, capture prey, and court their mates.

One of Lorenz's working techniques was the comparative study of similar behaviors in species known to be related, i.e. having their origins in a common ancestor through anatomical evidence. Figure 10.1 summarizes his interpretation of the FAPs of ducks and geese (Family Anatinae) in terms of homology and evolution. Lorenz noted that the entire group had a few FAPs in common. All showed "displacement piping of lost chick" and "displacement drinking as a signal of peace". He placed these behaviors at the base of his model. As one progresses up the angled lines representing species, the horizontal lines representing FAPs in common extend over fewer and fewer species. The arrangement of the angled species lines demonstrates which species have many or few FAPs in common with others. Lorenz also noted those species with exceptional differentiation of the FAPs as a useful piece of information and those which had lost FAPs from their repertoire. By arranging and rearranging his species to maximize the effects shown he was able to bring to the center of his model those species with the greatest number of FAPs in common. The final arrangement was dubbed "Lorenz's shaving brush model". Thirty years ago it was a novel attempt at a multivariate presence/absence statistical analysis, a good deal ahead of its time, as population ecologists and numerical taxonomists then ignored the problems of multiple variables in ecological and classificatory systems which so vex them today. It provides a species order which shows reasonable agreement with the conventional classification.

There is other similar evidence of substantial behavioral similarity between animals which were already known to be so similar structurally that they had been classified by taxonomists as closely related. Much of the work is on birds, e.g. Dilger (1960) on lovebirds and Tinbergen (1960) on gulls. The behavioral information adds to the degree of similarity between the various species, which to a biologist can only be explained by evolution from a common ancestor.

An example of modern phylogenetic (cladistic) reasoning about homologous behaviors and their lines of descent lies within the social organization and sexual behavior of apes and humans. According to Martin and May (1982) the size of penis and testicles, plus some other sexual features of primates, relates to social organization. Figure 10.2 provides a cladogram showing that a large penis and large testicles are apomorphic (i.e. novel adaptations) for human and chimpanzee respectively, whereas small organs are the

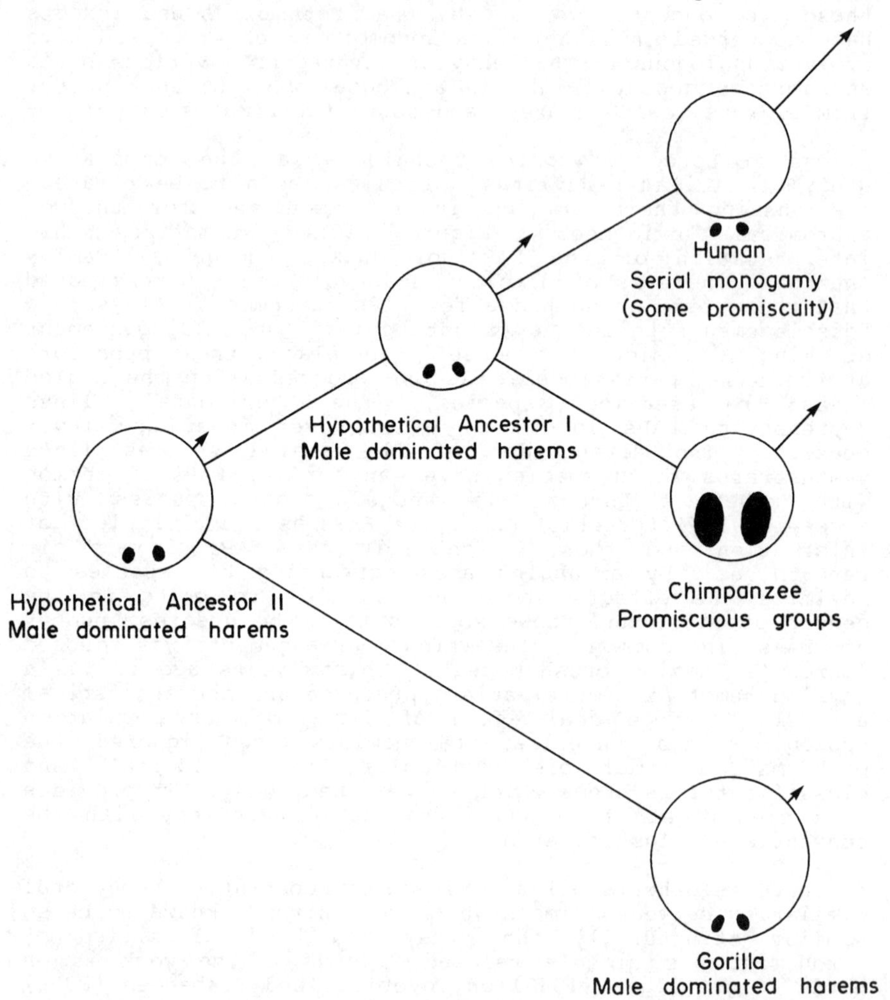

Figure 10.2. Cladogram of sexual behavior and social organization in the apes and humans.

symplesiomorphic (ancestral) state. Large testicles provide an adaptive advantage in promiscuous harem societies in that the male with the most sperm is most likely to fertilize any of the females. A large penis is postulated as being adaptive in monogamous bonding, and the disputable function of the adaptation does not really affect the phylogenetic argument. Small organs, as in the related gorilla and orangutan, reflect virtually uncontested male ownership of a group of females (in a harem and scattered solitarily, respectively -- see Chapter 11). The most parsimonious phylogeny for sex organ adaptations in chimpanzee and human is a common ancestor with a moderate sized penis and small testicles. This indicates that the common ancestor must have had a social organization which was not amiably promiscuous as in chimpanzees. Penile threat displays are a common primate behavior, so the ancestral moderate sized and visible penis suggests an overtly ranking society. The social organization of the ancestral form could have been one of a frequently contested rank structure between males and possessive control of estrus females by the dominant.

It is also noteworthy that Martin and May's diagram claiming that chimpanzee testicles are not descended into a scrotum as in man is easily disprovable by even cursory viewing of relevant movies or videotapes. According to the cladistic style of reasoning the common chimpanzee-human ancestor in Figure 10.2 would have had descended testicles, as would subsequent descendants on both human and chimpanzee lines. It is a homologous plesiomorphic character. The package of penis and scrotum surrounded in modern humans by pubic hair has the characteristics of a behavioral badge, flashable like a female duck's speculum. This type of phylogenetic reasoning suggests that penile threat displays are homologous behaviors in primates and appear in several lines of descent, including well along the human line. Needless to say, basing phylogenies and behavior on only two or even three structural characters can be misleading. The style of reasoning, though, has brought impressive advances in understanding phylogenies (Wiley 1981). It can be applied to structural/behavioral combinations so that we can explore the behavior of fossil species.

Natural selection

A second kind of evidence for the evolution of behavior is the experimental analysis of rates of change in relation to function. This is a much more critical, but also more difficult, approach because of the difficulty of establishing a measurable criterion for fitness. If the criterion is only expressed at the concept level of "being fit" it does not provide guidance on measurable parameters of what is "being fit".

One investigation managed to progress in measuring fitness, and has even started on the genetic analysis of some of the properties underlying the fitness. It is the investigation of industrial melanism (= darkening) in the peppered moth *Biston betularia* by Kettlewell (1973) and successors (Bishop and Cook 1975).

The peppered moth has an unusually high proportion of the normally rare dark form in heavily industrialized areas which suffer air pollution from soot and other dark smoke. Under such conditions trees, ground and buildings are covered with grime, and gradually darken. The peppered moth spends much time resting on surfaces during the day. In clean areas the light speckled form is cryptically colored when settled on light (lichen covered) tree trunks. In aerially polluted areas it is the dark form which is less easily visible due to the camouflaging effect of its melanism.

Kettlewell generated an ingenious field experiment to test whether or not the change which had occurred over about 100 years of industrialization in northwest England had survival benefits for the individual moths showing industrial melanism. He raised enough peppered moths of both light and dark varieties (from eggs, through caterpillar to adult stages - a long process) to be able to transplant adult moths onto light and dark trees in a clean wood in southern England, and in a dark wood in the north. After 24 hours of predation survival rates showed that light forms survived bird predation best in the clean areas, and dark forms survived best in the dark woods. By keeping close watch on the behaviors concerned he showed that the peppered moths would alight on the most suitably colored background or move over onto it, and spend many hours in daylight resting on that background. He also showed that bird predators would range tree trunks looking for prey, and could readily see the wrong form. They would hover, eyeing the prey, and then dart in to pick it off the tree trunk. The moth had a habitat-choice behavior (light or dark bark patches) which optimized the value of its protective coloring, and raised its survival chances (3:1 for cryptic as opposed to contrasting coloration) against sharp eyed bird predators.

The effect spread widely through England during the growth of coal burning industries in the 19th century, but is now receding due to aerial emission controls and the general lightening of the countryside from its early 20th century grime. The spread can be documented because of an army of amateur entomologists in the country. Their records have proven invaluable to the professionals and have added considerably to the documentation and analysis. This rapid spread of a species character over many generations is due to the increased survival of individuals bearing the character. It is natural selection in operation. In this case, the later retreat of the character resulted from reduced survival benefits as air pollution decreased in England, and the operation of natural selection against the character which, for a hundred years or so, had advantages in a restricted area.

As might be expected from the extent of the polluted countryside, peppered moths were not alone in their response to the darkening. More than 70 species of night-flying moths show industrial melanism, but they show different proportions of dark forms, and different distributions in the countryside (Bishop and Cook 1975). Some species are more clustered around the points of dark smoke, whereas others are more generally dispersed throughout the total area affected (Figure 10.3). There are species differences in response following species differences in behavior. These have not yet been well worked out.

Initially the dark form was believed to be a simple Mendelian recessive gene, but this concept of a simple mutation and rapid spread of a new quality throughout the population of the species had to be changed on close analysis. The peppered moth has a low level of dark forms in clean areas, but a higher level in parts of its range which are naturally gloomy, such as in the coniferous forests of northern Europe and Canada. There is even a range of different kinds of light and dark forms. Evidently there are balances affecting the expression of darkness which vary from area to area, and the dark character is available within the gene pool and can respond to the selection pressures of man-induced habitat darkening. Not only the dark character is available within the gene pool, but the corresponding behavioral characters of dark or light micro-habitat selection (bark color) are lying dormant there as well.

For industrial melanism to be adaptive both structure (body color) and behavior (dark/light bark choice) must work together. What is selected is a functioning balance between the two sets of characters. The balance in any habitat provides survival advantages in that habitat, and is quite responsive to environmental changes.

There is an illuminating interplay of inherited and learned behaviors, and how they may affect the fitness of an individual, among hybrid *Agapornis* lovebirds. These birds carry nest material either in their beaks or tucked between tail feathers. Hybrids (Dilger 1964) between species genetically programmed for the two different behaviors are ineffective at either method when first nesting but can eventually learn beak carrying, and may even learn tail-feather carrying. In this case their learning capability is able to function as a backup mechanism, overcoming their inadequate hybridized genetic programming (see Chapter 5).

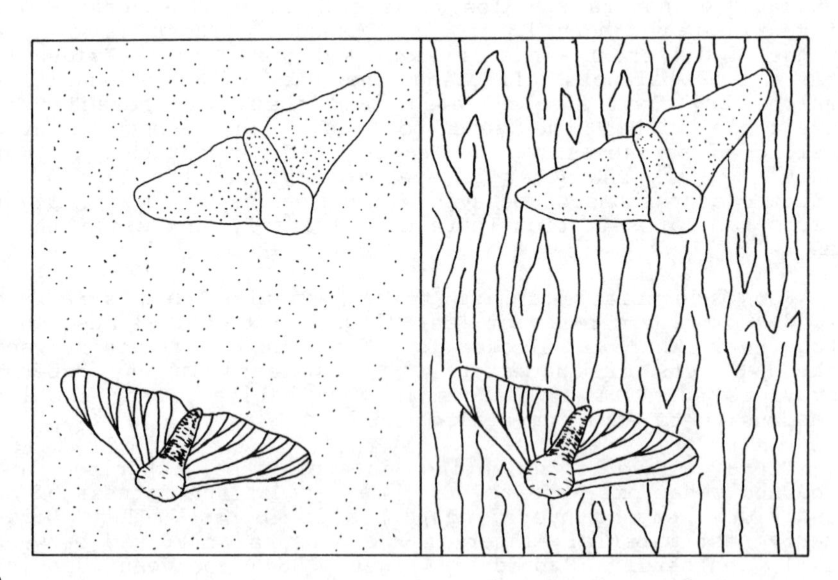

Figure 10.3. Location (a) and camouflage effect (b) of industrial melanism in the peppered moth. Black arcs of circles in (a) show proportions of melanistic individuals.

Allopatry, sympatry and vagility

There is a continuing controversy in evolutionary theory whether geographical separation between demes (allopatry) is essential for speciation. Geographical separation means physically (or chemically) impassable barriers such as mountain ranges or intervening seas. The alternatives to allopatry are forms of sympatric or parapatric speciation, with more or less complete or only partial range overlap respectively. A key component which can influence the potential for sympatric speciation is the behavioral property of vagility, the degree to which individuals can disperse through their species' total range. With high vagility, there can be much gene flow, and little chance of sympatric speciation.

Many species, however, are comprised of individuals which do not range freely through their potential habitat, even though physically capable of doing so. There can be a very limited gene flow indeed in active species with a really low vagility inherent in sets of behavioral adaptations such as territory, group home ranges, and rigid social organization determined by an old experienced dominant. Measures of actual, as opposed to potential, vagility for particular species is a needed contribution by behaviorists to arguments over allopatry and sympatry.

Sociobiology: its contribution to evolutionary theory

A number of evolutionary concepts developed during the 1970s centered on the basis that natural selection acted on individuals. Proponents of the concepts can be loosely grouped under the term sociobiologists. They were critical of the common way of expressing natural selection as a force operating on the species as a unit, or on a group of individuals. They reaffirmed Darwin's original argument that an adaptation of a species to particular ecosystem demands is the product of the survival of individuals showing the characters such that their improved survival spread the character throughout the species. When a species has group organization, then it can be that the beneficial character will first spread through a group, and then if the group survives better than other groups as a result of the character, this will speed up the spread of the character through the species.

In addition, the argument runs, characters which benefit the species but not the individual, such as Wynne-Edwards (1962) epideictic behavior (behaviors acting as sign stimulii for overpopulation resulting in a response of lowered fecundity), could not occur. The characters would be bred out of existence. Individuals not showing the reduced fecundity would pass on their high fecundity qualities at the expense of individuals with lowered fecundity.

From this base concept of survival of individual animals a number of others were derived.

Survival machines

To understand the genetic foundations of behavior and the very considerable degree of genetic similarity between organisms (Maynard Smith 1975) it can help to consider them as "survival machines" (Dawkins 1976). Organisms are simply machines which contribute to the survival of packages of genes.

The concept of survival machines is developed this way. Once complex strings of organic molecules floating in the environment developed the property of self-duplication, natural selection favored the spread of those strings which duplicated themselves quickest. Once the strings became so abundant that they had to compete for limited environmental chemical resources to form their duplicating strings, natural selection favored the survival and spread of those strings which processed the needed resources most effectively. Under these selective pressures, any molecular strings which could encase themselves in progressively more efficient (complex) bodies specializing in absorbing nutrients, or eating other bodies already containing materials in compact concentrated packages, had survival advantages. Molecular strings diversified, specialized and evolved to the DNA helices with component genes which determine the structure and behavioral strategies of an infinite array of living packages, each one of which is programmed by eons of selective pressures to deploy a variety of strategies to assist in its survival and pass on copies of its gene complement.

In behavioral terms animals are programmed by their genes to build whatever structures and perform whatever actions which will maintain them alive and allow them to breed. It does not matter to an animal's genes how it does it, provided that it does do it.

The concept of survival machines reinforces the idea that it is the individual animal that responds to selection pressure, and either passes on its genes or does not. Other consequences to the group or species follow from the survival of individual animals.

The concept has been misunderstood in many ways ranging from thinking of genes as sinister controllers of animal and human destiny, to inflexible determinants precluding learning and intelligence in animals and humans. Learning and intelligence are two of the strategies open to survival machines, and if learning and intelligence can best do the job of keeping a set of genes perpetuating themselves, then those genes which interact to produce those characters will survive.

The big evolutionary jumps

There is a tendency to think of selection pressures in terms of today's world of competing, highly specialized herbivorous, carnivorous, and parasitic species, and superbly efficient opportunists. Recent postulates that in pre-Ordovician times the global marine ecosystem was occupied by organisms so diversely structured that present classification would type them as belonging to many more phyla than now occur, suggest that there were very different selection pressures at that time. For so many phyla to coexist it must have been possible for major structural changes to occur relatively easily. For this to happen there is the implication that at that time animal activities were performed so inefficiently compared to today, that there was scope for inefficient animals to get by and survive, and associated ontogeny so poorly organized that neotenies could survive even with inefficient survival and breeding behaviors.

Once active and large animals appeared and evolved the strategy of living in organized groups (as in the Ordovician period), a new dimension of selective pressures appeared which we see today, at the expense of understanding what such pressures were like previously. Modern evolutionary concepts are largely based on modern ecosystems and their processes, and some of them may not be relevant to pre-Ordovician times. One example is that the original evolution of metazoa from pre-existing Eukaryote cells must have occurred in an ecosystem lacking large predators and suspension feeders. Another example is that the present correlation between planktotrophic larval development, large body size (permitting accumulation of gametocytes) and mutually stimulated mass spawning, would not be needed in early ecosystems lacking efficient larval eating suspension feeders. Thus a larval strategy could have been available to all early sessile metazoa.

Inclusive fitness

When an animal acts in such a way that it dies but two of its sibs live, then the same number of genes that it carries survive since sibs carry 50% of each others' genes. For the genes concerned just as many survive whether they were in one survival machine or its two sibs. Accordingly, by chance genes could appear which program their survival machines to be altruistic. As a result of lethally altruistic behavior, the gene complement within the self-sacrificing animal may survive but within the bodies of a minimum of two other sibs, or four first cousins, and so on. The fitness of the individual altruist may seem low, but it is superseded by the high value of what is termed its inclusive fitness, the fitness of its gene complement as included among its sibs, its cousins, and other more distant relatives. This gene survival strategy reduces the conflict of interest in group living. Animals can live in groups, competing for resources to some extent, but if the group comprises kin, especially closely related kin, altruistic actions of one which contribute to the survival of others add to the fitness of the gene pool rather than the fitness of the individual animal.

Even the most social of birds and mammals do not live only in kin groups of sibs, since the advantages of altruism must be balanced against the costs of conflict with other behaviors, such as the genetic consequences of incest and reduced level of gene exchange between related sexually reproducing animals. Like many other adaptive qualities, the benefits and costs are reckoned statistically. The net benefits of inclusive fitness are that social groups will normally comprise related individuals, but will have adaptations allowing some gene exchange between unrelated groups. There are many strategies for making the exchanges; some are species-specific. Thus the gene exchange may be by females changing groups, usually when they become independent young adults, as in beavers, African hunting dogs and chimpanzees. Or the males may change groups, as in lions, baboons and elephants. Or groups may split when they are too large as elephants and killer whales do. Other species may receive no benefit from a formal strategy and group composition may change in many informal ways. The essential components are that the potential for altruistic behavior is balanced against other potentials such as gene exchange. The more closely related group members are, the higher is the inclusive fitness of individual behaviors which promote the survival of the group as a social colony, the more likely that the group will survive competition with other groups for limited resources, and the more likely the individual will survive due to the benefits of reciprocal or kin altruism within the group.

The hymenopteran insects, the ants, bees and wasps, have chanced upon a genetic composition which opened up a unique scale of inclusive fitness, altruism and social organization through the division of labor into castes.

In these insects males are haploid and have only one set of chromosomes and genes. Females are diploid, however, and have the normal two sets of chromosomes and genes. As a result, females receive half of their genes from the double set of their mothers and the other half from the single set

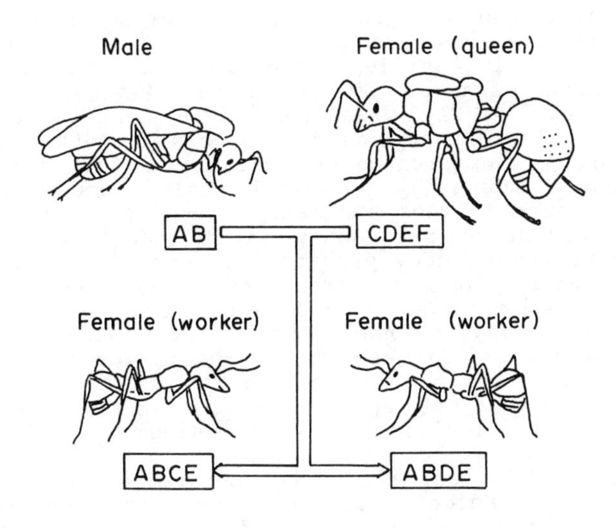

Figure 10.4. Gene composition of Hymenoptera.

of their fathers (Figure 10.4). In these societies, the most caste organized species reproduce from a single queen fertilized by a single male. The maximum genetic relationship achievable is that sisters carry 75% of each others' genes. If more than one queen breeds or more than one male fertilizes one queen, then the genetic similarity between females is less, but still more than the 50% of sisters in fully diploid species. Such females are more related to their sisters than their mother, the queen, is to them or than they would be to their own brood. The inclusive fitness of altruistic behavior which helps sisters to survive is greater than the individual fitness of breeding. This statistical complex has produced a variety of social colonial behaviors and caste evolution, including the honey bee pattern of female workers with food signalling dances, one-shot self-immolating defensive sting hardware, and one-shot fertilization from competing drone males with no other function within the group.

Some diploid animal groups have converged towards similar caste organized colonies, but lacking the haploid-derived high genetic similarities they have had to ensure very reduced gene exchange by us-and-them warring behaviors. Termites have highly organized caste based colonies and strongly resist intrusions by members of other colonies. Gene exchange can presumably take place during occasional simultaneous swarming in which males and females of different colonies meet and mate. Termites commonly retain multiple breeding, presumably as an adaptive balance to their edibility to many predators (compared to Hymenoptera) and the high rate of predation when they emerge as defenseless and tasty food from the previously all protecting colony. Once again, the potential of one quality, inclusive fitness through social organization, has to be balanced against the potential of other adaptations, the whole complex making for a viable species through the viability of its individuals (or colonies).

A few other animals have converged on the caste organization of colonial living. These include the African mole rat, the only mammal to adopt the strategy. This species has a breeding female, drone (stud) males, worker females of several types, and defense behavior against conspecifics from other colonies. Colonies are far apart and successful intrusions rare (although they must occur).

Some birds and mammals are close to having adopted a colonial strategy. They have not, however, separated into true anatomically distinguished castes adapted to specific work duties. What is called "aunting" and "uncleing" behavior is fairly frequent; non-breeders assist in the care of colony young. This behavior often occurs among pre-breeders, in other words the young uncles and aunts may eventually breed themselves. The early assistance period also has the advantage in mammals of young adults learning parental behavior. In some cases, such as that of wolves, subordinates may not breed at all and are constrained to the inclusive fitness behavior of uncleing and aunting, or they may breed less frequently and with less success than dominants. The dominants may even destroy the brood of subordinates. In short, there are a variety of behaviors adapted to balancing how the potential of inclusive fitness is actually adopted by a species.

Evolutionary Stable Strategies (ESS)

The quantitative approach followed by some authors (Maynard Smith 1975, 1978) has produced the novel concept of evolutionary stable strategies (ESS). The process of evolution will produce adaptive strategies which are stable in the sense that if the selective pressures remain constant, any deviation in another generation will be less well adapted and the individual less likely to survive and breed. This seems intuitively reasonable.

Aggression between group members balances competition for resources but carries the risk of being disabled, or disabling kin carrying similar genes. The ESS here is a balance between threat displays and the threshold to attack if the displays do not work. Depending on the species-specific balances, such as possession of a lethal armory of fangs and claws, in any one species, it may pay to always threaten, and have a high threshold to attack. In other species without lethal weapons the optimum balance to ensure individual fitness may be to always attack, without any threat.

There are extensions to this concept (Dawkin 1976). Since much behavior is social, an ESS may have a counter ESS practiced by the recipient of the social action. Thus within a species, aggressors (Hawks) may come into balance with a proportion of submitters (Doves) since the winning aggressor may damage itself whereas the withdrawing Dove survives unharmed. Games theory has produced many models to show balances between such strategies as Hawks, Mice (= Doves), Bully (initially Hawk, but changes to Mouse), Retaliator (initially Mouse, but changes to Hawk) and Prober-Retaliator (who tests carefully before changing to Hawk). The models can be made to show that such a population might evolve mainly to Retaliators and Prober-Retaliators and a few Mice. In other words, a strategy of coming on defensively but monitoring for competitor weakness is better than programmed aggresiveness or submission.

The models develop intriguing behavioral strategies, one of which is Cheat. In species where Cheating is a stable strategy (ESS), the Cheat uses a minority tactic for an adaptive behavior, e.g. spawning. Thus the top subordinate male in a group of spawning salmon may cheat at the last moment and quickly drop milt into the redd when the dominant is fully engaged in his extromission. There can be substantial survival advantages in cheating (Dawkin 1976), but only to a limited extent since once naive individuals are eliminated, Cheats would be left only with other Cheats.

But the Cheat cannot cheat without a dominant male going through the full courtship sequence.

Summary

To a biologist the ultimate cause of behavior is evolution operating through the twin forces of genetic change and natural selection. Genetic change provides the potential for new behaviors based on structures and physiology. Natural selection proves which new behaviors can work by bringing them into the balance of characters that allow the animal to survive and breed, i.e. to be fit. The existence of homologous FAPs in species shown to be related by conventional anatomically based phylogenies, is evidence for the evolution of behavior. Evidence also comes from the experimental demonstration of natural selection determining structural and behavioral changes over generations. The peppered moth and its camouflage behavior adapting to industrial melanism is an outstanding example. The survival percentage of light and grey moths in industrial and clean areas was the measure of fitness. Measures of vagility are important contributions by behaviorists to speciation arguments.

There are a number of novel and intriguing concepts in modern evolutionary theory as it is applied to behavior. Animals may be considered to be survival machines for the genes that they carry in the sense that the genes will achieve necessary adaptations by whatever structures and behavioral strategies are available to them. If there are only limited and convergent tactics for achieving a strategy, those tactics can be produced by many different gene combinations, as in the flying behaviors of insects and birds, and the courtship ensuring the mating of many animals. The individual fitness of an animal is extended by its inclusive fitness, i.e. the extent to which its actions assist in the survival of other individuals bearing similar genes. The extreme extension of inclusive fitness occurs in groups of very closely related kin organized into castes by the division of labor. In such species as ants, bees and wasps, altruistic behavior can raise the inclusive fitness of an individual by contributing to the survival of kin. Some behaviors and complexes of behavior are evolutionary stable strategies (ESSs) in that they represent the best balance of adaptations and cannot be improved upon provided the selective pressures remain constant. The concept does not imply that a single behavior is the best, but that a balance between one or more options may be optimal for individuals, especially if they live in organized groups. Thus a small proportion of Cheats may survive within a population of Non-Cheats, or a balance between Hawks (aggressors) and Doves (submitters) may be stable.

CHAPTER 11. THE BEHAVIOR OF MONKEYS AND APES.

One of the continuing questions of animal behavior is to what extent it is relevant to understanding the behavior of humans. Whether it has any relevance or not, we can be sure that the answer must be based on some knowledge of the evolution of humans and the species which most closely share with them a common line of descent. These close relatives of ours are the apes, monkeys and the prosimians, of the order Primates. The primates in turn are a mammalian order similar in taxonomic rank to the carnivores, rodents, cetaceans and so on.

We must have some knowledge of the primates and their fossil species so that we can critically appraise the concept of a common evolutionary origin for human and animal behavior by a reasoned line of descent, each animal in which was able to survive and breed within a real ecosystem which existed while it was alive. Arguing without that knowledge is a waste of time.

Since it is a question of "understanding", the answers can be expected to come in terms of proximal and ultimate causes. Some of the answers, particularly those concerning proximal causes (of physiology), are simple and the information readily available in texts on anatomy and physiology. Other answers, largely those concerning ultimate causes, can come from two sources: from observations on the behavior of closely related species, and from the structure of fossils.

Fortunately, some 20-30 years ago several groups of animal behaviorists, committed to the principles of what we can loosely call the school of ethology, turned their attention to studies of primates in the wild (e.g. Goodall 1971 and Kortlandt 1972). They followed the trail of a few innovative scientists and naturalists of the 1920s and 1930s, such as the American Carpenter (1934) and earlier yet (1900-20) the almost unrecognized South African Marais (1969). They substantially changed the approaches of urban biologists who had made what they could out of their limited opportunities (Zuckerman 1932) of working with over-populated, sex-mad, zoo colonies. Simultaneously in the 1940s and 1950s experimental psychologists progressed in understanding the complexities of the primate mind relative to human style emotions and breakdowns (Harlow and Harlow 1962). Infant monkeys deprived of real and surrogate mothers were pathologically withdrawn and as adults abusive incompetent parents.

We now have a body of information which allows us to partition tentatively similarities between humans and other primates into homologies and analogies. With corresponding fossil discoveries and taxonomic ordering, we can inductively develop rational even if conflicting phylogenies for the primates (e.g. Passingham 1982 and Gribbin and Cherfas 1982). These indicate when and how a human line of descent became one of those radiating away from the base arboreal stock and adapting to characters favoring survival away from trees and their canopies.

Structural and physiological similarities

The majority of primate species are arboreal. They have structures adapted to three-dimensional living within a shaky habitat of interlocking plant filaments, strings and rods, only a few of which are of substantial size and strength. These last (tree trunks) tend to be vertical rather than horizontal. Primate behaviors are adaptations to this habitat, where vision and olfaction are often limited by the canopy of leaves to relatively short distances, but sound transmits a long way. A three-dimensional life was a novelty to the prior ground dwelling almost two-dimensionally living mammals, and there were survival advantages to those animals which could evolve a number of new adaptations. Some of these must have been achieved very early in the evolution of the stock since they are held by even the most primitive forms.

The limbs are jointed in ways which facilitate their extension through three dimensions rather than being largely limited to two-dimensionally articulating levers for running. The limbs terminate in the grasping devices of hands and feet rather than weapons of claws or leverage pads of paws and hoofs. The digits are extendable like the arms and legs, facilitate grasping, and have a backing of firm, resistant plates (nails).

The eyes face forward, and the muzzle is shortened or dropped, permitting three-dimensional vision, directed forward. Mobility of the head sensors is due to flexibility of the neck, or by directing the attention of the whole body. The retina is provided with ample cones; the eyes allow near focussing. There is a well developed visual cortex in the brain, and the animals draw on much of the potential of the visual sense. Olfaction, in contrast, is reduced in importance although a few species, especially ground or trunk users, retain a sensitivity and scent glands for marking purposes. A thick forest can give little guidance to the direction from which pheromones are being released. The skeleton is structured with modifications of the backbone, pelvis, limbs and skull so that many species have the ability to sit, stand, climb and walk upright. This extension of forward and turning movement into a third vertical dimension of space facilitates activities in a network type habitat.

Taxonomy, habitat and distribution

The formal taxonomy of the order is shown in Table 11.1, with notes on habitat occupation, distribution and population trends.

The prosimians are generally small and represent a sequence of adaptations to arboreal life ranging from tree shrews, which are not very different from the simplest order of placental mammals, the insectivores (shrews). Some species have become nocturnal with greatly enlarged eyes (e.g., the Tarsier). Others (e.g., the Indri) have become large, and were forced to return to partially living on the ground or remaining close to central trunks and branches. The chances of continental drift separated the island of Madagascar from the African mainland before the evolution of more advanced primates (Figures 11.1 and 11.2). The result is that the prosimians have undergone an adaptive radiation in Madagascar and occupy niches there no longer available to them where they are in competition with other forest dwelling species. Other than Madagascar, they are limited to Africa and S.E. Asia.

The second of the two suborders, the Anthropoidea, includes all the monkeys, but also apes and humans. There are three divisions of the suborder, each ranked as a superfamily. They are the New World monkeys, the Old World monkeys, and apes and humans. (For the formal superfamily names see Table 11.1.)

The New World monkeys are arboreal. They show little tendency to grow very large or revert to ground living. They have special adaptations which are either not present in the Old World forms, or are extensions of simpler versions. The tail is prehensile and functions as a fifth supporting appendage. It greatly assists the New World monkeys in working their three-dimensional space, especially in behaviors involving suspension, by freeing up a digit bearing paired appendage for grasping and holding for inspection. Vocalization may be specialized for long-distance calling, and the dawn group roaring of howler

Table 11.1. Taxonomy, habitat and distribution of the order Primates.

Order - Primates.

Suborder - Prosimii. Lemurs, lorises and tarsiers, etc. Small forest dwelling forms with arboreal adaptations. Africa and Southeast Asia. The Island of Madagascar has shown an adaptive radiation of these forms.

Suborder - Anthropoidea. Monkeys, apes and humans.

Superfamily - Ceboidea. New World (South and Central America) monkeys. Forest dwelling forms with minor differences from Old World forms including prehensile tails. Many pronounced vocalizers.

Superfamily - Cercopithedae. Old World monkeys. Forest dwelling forms generally, but include vervets, macaques and baboons, forms which show a sequence of adaptations to ground living.

Superfamily - Hominoidea. Apes and humans.

Family - Pongidae. Apes. Gibbons and siamangs are arboreal, as are orangutans. They occur in Southeast Asia or adjacent East Indian Islands. Gorillas and chimpanzees are ground dwelling and inhabit isolated, recently substantially reduced, ranges in Central Africa with slightly different forms.

Family - Hominidae. Humans. Humans are ground dwelling with worldwide distribution. Recent population explosions and emigrations have increased numbers and interbred different forms.

monkeys can be heard by many separated clans (Carpenter 1934). The New World ancestral stock is believed to have occupied the western area of Gondwanaland at the time of the African-South American separation and continental drift apart (Figures 11.1 and 11.2). The limits to distribution now are the forests of South and Central America.

The Old World monkeys have spread from the common African source eastward through the forests of Asia (Figure 11.3). Within the subfamily Cercopithecinae they have also generated a series of forms of increasing size, corresponding to re-adoption of ground living habits. The

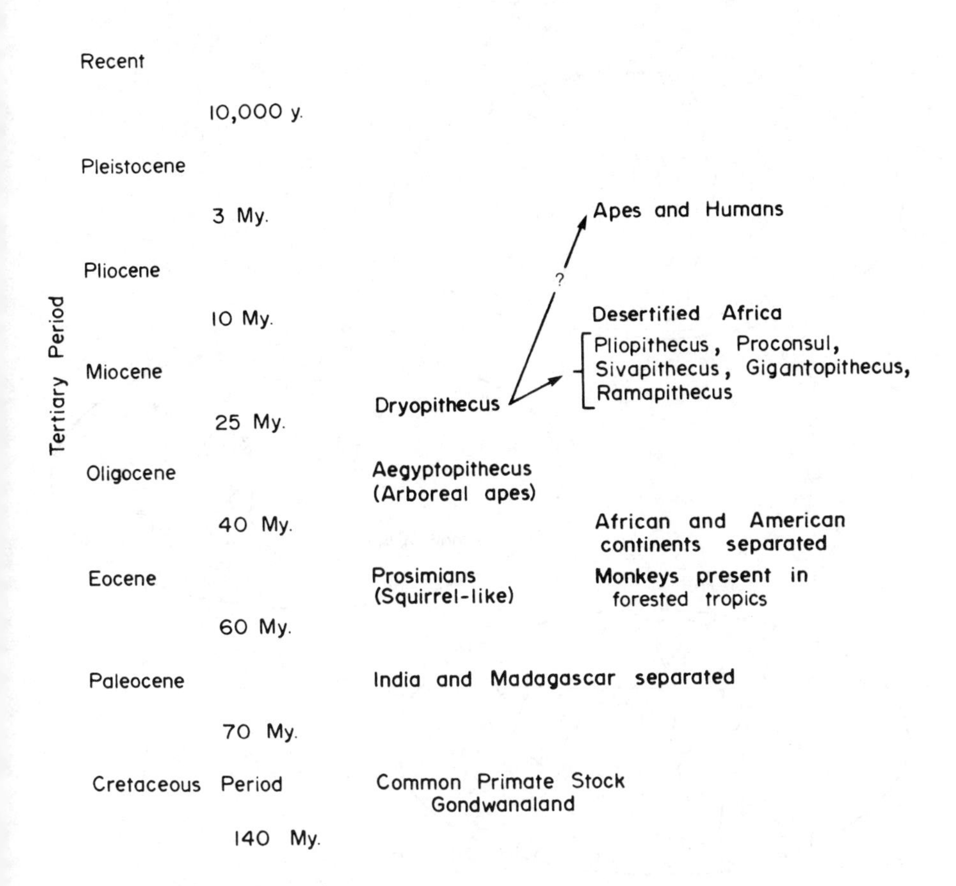

Figure 11.1. Geological chronology important in the evolution of the primates.

ground living forms have outdistanced the arboreal forms in their spread. The northern barbary ape of Gibraltar and the snow monkey of Japan are both ground dwelling macaques.

The smallest in this series, the vervets, are occasional ground dwellers. They are small, of customary monkey size. They can occupy a habitat of bushlike trees marginal to savannah, grassland, lake, lagoon or sea, and by daily ranging of meadows, wetlands and beaches adjacent to their safe canopy roosts, increase the diversity of their resources. Such an initial adaptation facilitated the eventual spread of ground dwellers away from the virtually

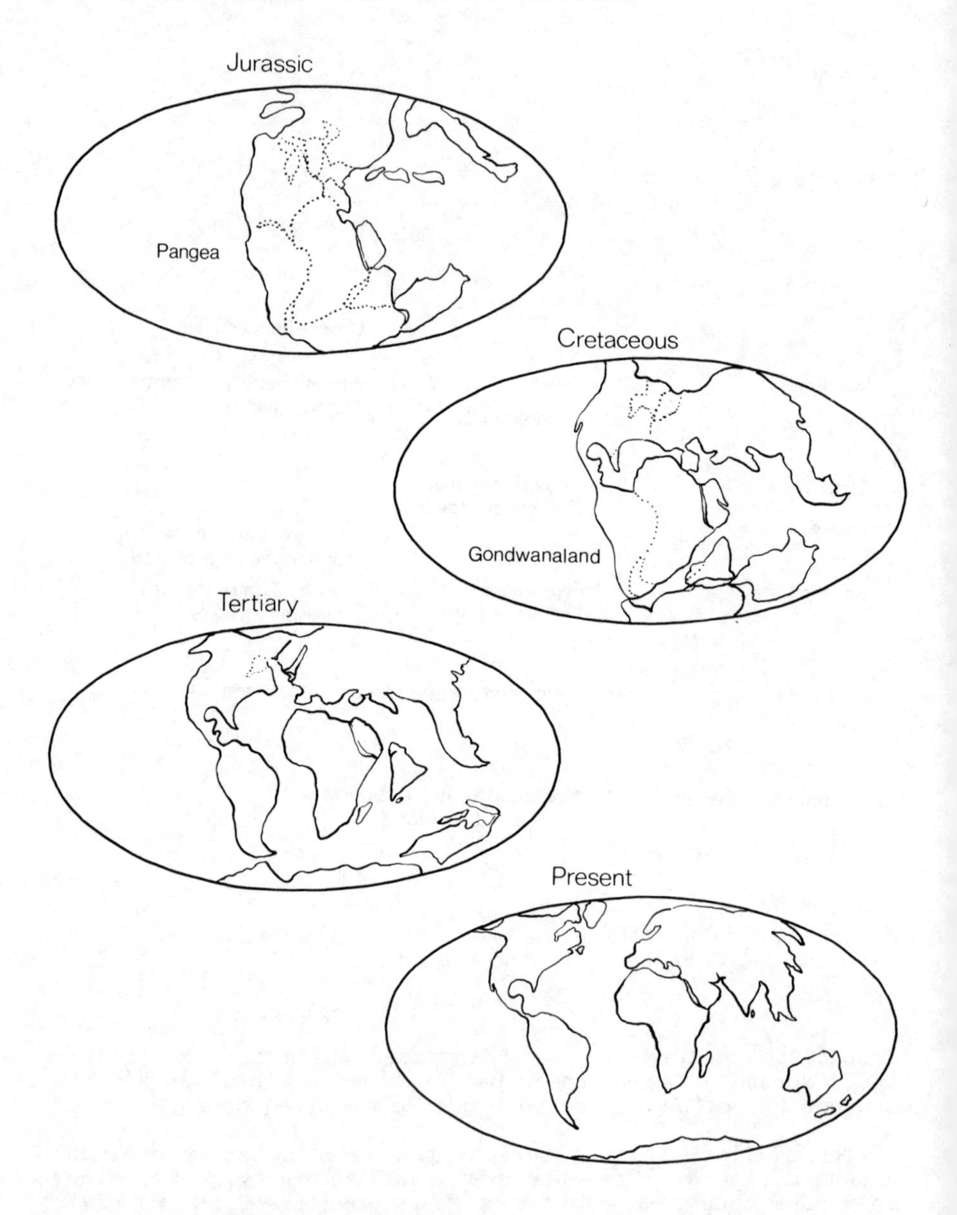

Figure 11.2. Continental drift relevant to understanding present day primate distribution.

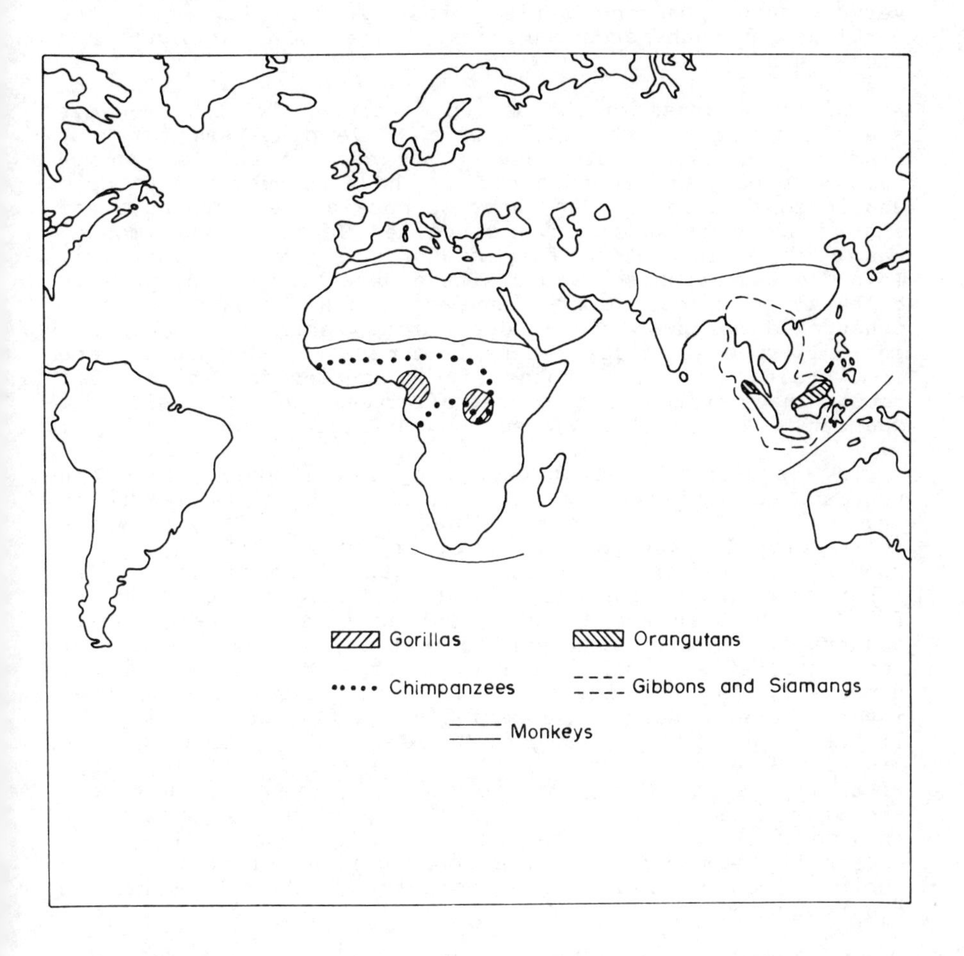

Figure 11.3. Distribution of Old World monkeys and the great apes.

seasonless tropical forests into ecosystems with seasonally productive and unproductive periods such as winter and summer, or monsoon and dry. Grassland shows extremes of production but also dry season die-back. Wetlands show delayed drying out during drought periods whether they are seasonal or irregular. Beaches provide resources of scavengable flotsam and dessication resistant shellfish.

There are a variety of forest-verging habitats but the vervet-macaque-baboon series has largely entered the parkland-savannah-scrub habitats, rather than wetlands and shorelines.

The next larger in the series of ground dwelling monkeys are the macaques, the forms which have dispersed farthest from the ancestral grounds in Africa. Macaques can be almost completely ground dwelling, have accomodated to cold and seasonal climates, and have spread into eastern Asia and its offshore islands. The so-called Japanese snow monkey has a thick insulating fur, assisting in winter survival. It and other Japanese macaques have developed group-specific cultural practices, some based on beach living. Species penetrated northward to Europe during warmer climates, but the only species still present is a macaque, the barbary ape of Gibraltar, a fully ground living species for which cliff roosts have taken the place of trees as safe sleeping quarters. The stock now needs conservation.

The largest of the monkeys are the baboons. They are represented by five or six species (there is taxonomic argument) separated into two genera (Table 11.2). Some are fully ground living, having substituted cliff for tree roosts as sleeping quarters like the barbary ape. The others are ground rangers, spending daylight hours looking for food between known safe roosting trees. Their habits constrain a number of their activities, in the sense that roosts have a limited carrying capacity for sleepers, especially for a species based on very aggressive social interactions. Roosts, especially cliff roosts, may be limited in number and clustered, thus limiting the distribution of roost dependent species and creating virtually genetically isolated demes. A few baboon troops are shoreline rangers (Davidge 1976). Thus all groups in the series vervets, macaques and baboons show that the seasonally dependent resource opportunities of their ground habitats can be realized especially if they live near a coast.

The third superfamily, the Hominoidea, is divided into three families, the Hylobatidae (gibbons and siamangs), the Pongidae (the great apes) and the Hominidae (humans). The Pongidae includes three groups of species, each with several closely related forms, which are not separated at the species level (Table 11.3).

The Asiatic gibbons and siamangs are the smallest and most arboreal of the apes. They have some difficulty in running on the ground because of their unusual body balance arising from long arms which allow them to brachiate through their forest canopies. They are extremely agile in the trees and spend most of their time there.

Table 11.2. Taxonomy and habits of the baboons.

Scientific name	Common name	Comments
Papio hamadryas	Hamadryas (sacred) baboon	One-male groups within bands. Bands can share sleeping cliffs. Dry country.
Papio anubis	Olive (anubis) baboon	Multi-male groups with males capable of ranking alliances.
Papio cynocephalus	Yellow baboon	Parkland and savannah.
Papio ursinus	Chacma baboon	
Theropithecus gelada	Gelada	One-male groups within bands which can aggregate. Dry country.

The orangutans are large, solitary and slow. They are capable on the ground, and when roaming usually do so on foot. However, they spend much of their time feeding slowly in the canopy. They occur on the southeast Asian islands of Sumatra and Borneo.

The other two groups are African and ground dwelling. The gorillas, both mountain and lowland forms, are so big that only the largest and strongest trees can support them. They may roost in trees, constructing nests of bent over boughs which are occupied individually for one night only and then abandoned as the group moves on. The chimpanzee, including the pygmy form, gathers food equally well in trees or on the ground. It socializes on the ground, making extensive use of clearings, although the male displays may extend into the tree canopy. Chimpanzees may also hunt occasionally using socially organized combined ambush and run-down tactics in which one or more chases a monkey or other prey through trees and bushes, and the remainder of the group waits below or in a strategic position to intercept.

Humans are essentially ground dwelling, although capable of building artificial ground stacked in vertically arranged habitat modules of apartment blocks, office towers and other

Table 11.3. Taxonomy of the extant apes.

Taxon	Common name
Family Hylobatidae (S.E. Asia)	
Hylobates	Gibbon
H. lar, H. agilis, H. moloch, H. concolor, H. hoolock, H. klossi	
Symphalangus	Siamang
S. syndactylus	
Family Pongidae	
Pongo	Orangutan (S.E. Asia)
P. pygmaeus	
Pan	Chimpanzee (Africa)
P. troglodytes troglodytes	- central chimpanzee
P. t. verus	- western chimpanzee
P. t. schweinfurthis	- eastern chimpanzee
P. t. paniscus	- pygmy chimpanzee
Gorilla	Gorilla (Africa)
G. gorilla gorilla	- western lowland gorilla
G. g. manyema	- eastern lowland gorilla
G. g. beringei	- mountain gorilla
Family Hominidae (worldwide)	
Homo	Human
(*H. sapiens*)	

constructions. The key habitat adaptation is that they have developed manipulating capability allowing them to distribute themselves wherever their life-support habitats can be constructed. By such manipulation they substantially control their ecosystem interactions with predators, parasites (pathogens), prey and competitors, although subject to unexpected ecosystem responses generating plagues, pollution and war.

Primate fossils and continental drift

A variety of fossil forms appear in the geological record, and their timing must be put together with the timing of continental drift, to explain as far as practical the present day distribution and adaptations of prosimians in Madagascar, New World monkeys in South and Central America, Old World monkeys in Africa and Southern Asia, and apes in tropical Africa and Southeast Asia and offshore islands (see Figure 11.3). There are some interesting gaps and inconsistencies in the literature.

The initial split of land mass Pangaea (Figure 11.2) which gave rise to the present day continents is generally dated at about 150-200 million years (Myrs) ago in the late Jurassic. Madagascar (with its prosimians) separated from Gondwanaland (or the east coast of what is now Africa) about the early Cretaceous (Figure 11.1), 140 Myrs ago (Norton et al. 1979, Audley-Charles et al. 1981). The split between Africa and South America with their respective New World and Old World stocks is generally believed to have started a little later in the Cretaceous, say 100 Myrs ago, with the rift developing at the south and by rotation progressing over the millions of years to the western bulge of Africa, which is now the Gold, Ivory and Grain coasts. This timing is sufficiently earlier than the appearance of prosimians in the fossil record (Paleocene and Eocene - about 60 Myrs ago) to leave some uncertainty in the literature about the pattern of origin of New and Old World monkeys. The later the split was complete, the easier it is to accept that Old and New World monkeys came from a joint monkey stock evolved from the prosimians. Passingham (1982) claims an Eocene split (40 Myrs ago) and Gribbin and Cherfas (1982) a South American-African split similarly late (40 Myrs ago). The earlier the split was completed (if say, 100 Myrs ago), the more likely New and Old World monkeys had separate origins, presumably from prosimian stocks.

There are, for instance, fossil prosimians, apparently squirrel-like forms of the genera *Plesiodapis* and *Smilodectes*, about 50 Myrs ago, and Simons (1964) considers that these could indicate a stock present in the Cretaceous (70-140 Myrs ago) at the time of the continental splitting. If so, a common stock can account for prosimians in Madagascar (never any competition with the more efficient monkeys later evolving from the prosimian stock drifting off with the South American and African continents) and the distinctive New and Old World monkeys in South America and Africa. The latter dispersed into Asia after Eurasia connected up with Africa, and presumably eliminated by competition almost all prosimians that drifted along with India until connected with Eurasia. Australia and Antarctica apparently split off Pangaea too early for prosimians, or placental mammals in general. It simply has to be accepted that there are enormous gaps in the fossil record, and some uncertainties in relative timing of continental drift and fossil primates.

The next set of significant fossils are those of believed arboreal apes: *Aegyptopithecus* (28 Myrs ago) and the Dryopithecines (20 Myrs ago). The latter are *Pliopithecus*, *Proconsul*, *Sivapithecus*, *Gigantopithecus* and *Ramapithecus*, extending over the period 20-10 Myrs ago. Of these, *Ramapithecus* (14 Myrs ago) shows tooth cusps, jaw and bone shapes most strongly suggesting ape-like characters and possibly habits - particularly brachiation. Then comes another gap in the record through the hot Pliocene of 10-4 Myrs ago, which is significant for understanding the evolution of humans and human-like forms (see Chapter 12).

Behavior

The two main behavioral qualities which distinguish the primates from other mammals are their level of social organization and their ability to learn and reason.

Social organization

Table 11.4 (modified from Chalmers 1979) lists species and the size of their groups, which is a property limiting the potential of the species for social organization. Chalmers starts his list with a warning that some of the data is based on single encounters of naturalists with the species and thus may be inaccurate. However, his arrangement of species into four sets indicates that there have been relatively few ways in which primates organize themselves into groups and that they have repeatedly converged on similar adaptations, whether prosimian or Anthropoid.

Chalmers' first set is that of solitaries. Included are species in which females and males are usually found on their own, with the exception of the young being with their mothers. This does not mean that there is no social interaction at all, since obviously males must find females on occasion. There is the possibility of territoriality with spacing of the animals, and knowledge of who is one's neighbor. There is also the possibility of separate male and female spacing.

Such spatial arrangements are known to exist in the mouse lemur species *Microcebus murinus* (Charles-Dominique and Martin 1972). Females all have their home ranges, which

Table 11.4. Social organization of primates.

Organization type	Group size	Home range size (hectares)
Solitaries	1	ca. 1 or less
Several lemurs and lorises		
Orangutan Male especially Female with young to 4-5 years Juveniles may group		
Monogamous family groups	2+	4-50
Breeding male, female and young		
Indri, some monkeys Gibbon, siamang		
One-male, several female groups	10-20	10-5000
Many monkeys Gorilla (may tolerate some mature males) Hamadryas and gelada baboons (groups, may aggregate)		
Multi-male, multi-female groups	10-100	1-2500
Some lemurs Many monkeys, including macaques and baboons Chimpanzee - frequent subgrouping, between kin and "friends"		

overlap so that they can keep each other company or be hostile on occasions. A single male, however, overlaps several females. Other non-breeding males occur (again solitarily) at the fringes of the territories of breeding males.

One of the apes is almost solitary in the same way. The orangutan is usually found on its own, other than the young up to four or five years old who are with their mothers. Adolescents, though, may group up.

Chalmers' "solitaries" is a term which means that much of the time individuals are spaced out sufficiently that their social organization is not obvious except to an observer who spends considerable time noting who does actually meet whom, how often, and where. The "solitaries" should probably be considered a set of species the individuals of which are dispersed so that the social organization must be played out over distances fairly large relative to the size and separated spacing of the animals.

The second set consists of monogamous species in which the predominant social grouping consists of a male and a female, with or without dependent young. Within this group the small apes, the gibbons and siamangs are aggressive in defending their territory from other adults, and eventually drive away their own developing juveniles.

Groups of one male and several females are widespread among the monkeys. The surplus males generally live near one of the breeding groups, either on their own or in bachelor groups. The males generally are much larger than the females, and strongly defend and dominate their female property.

Two of the baboon species, the hamadryas and the gelada, have developed this basic organization of one male with several females into a hierarchical organization. Groups live together in bands, individual males form alliances or friendships within the bands at least for grooming and associating purposes, and the band may function as an entity in fighting other bands. The band has a range and may associate with other ranging bands to share roosting cliffs, on which at any one night several hundred individuals may be gathered. They make a terrible racket.

One of the apes comes into this category. The gorilla normally is found in small groups of this type, although on occasion more than one male has been seen in a group. The groups are organized with little violence, although there may be spectacular chest drumming and calling displays. Relationships between males can be so amiable that a subordinate male has been observed actually mating in full view of the dominant. Thus the sexually possessive term "harem" is not necessarily appropriate for this kind of group.

The final set of species with multiple male groups spans all the superfamilies - lemurs, New and Old World monkeys and one of the apes. It could be a rather artificial group comprising species in which males live tolerantly together, those with a fierce rank order which determines the male's social position and alliances, or even those in which the tolerance of the dominant male to growing young is extended until they, as young adults, show sexual interest in the females.

Once again the baboons show organizational developments which go beyond those of other monkeys. In the multiple male baboon groups, males may make alliances and thereby dominate the social organization. Each individual of the alliance may have to defer to other stronger, often younger males when away from his allies, but he takes care to keep himself strategically placed so that he can call on them during aggressive encounters.

There is yet another organizational novelty among the ground dwelling macaques. Alliances of older females can moderate the dominance of the larger males, and even inhibit it. These groups come close to being matriarchal societies.

The one ape in this set is the chimpanzee, another ground dwelling species with novel social organization and arrangements. Chimpanzees live in groups which may number up to about 80 individuals, but the groups are flexible. Subgroups may break off, particularly for the day during times of food shortage (sometimes for longer periods), but they reassemble with greeting ceremonies and obvious recognition. Particular individuals may form long-term associations. Some of these associations are known to be between sibs, particularly male-male ones, but they also may be formed between mother-daughter and mother-son. Group splitting can be long-term, even permanent, and on at least one occasion the split was followed by group hostility, and the lethal raiding of one of the split groups into the other's territory. In comparison with baboons, though, the multi-male groups of chimpanzees are peaceful societies, with males given to occasional leaping and calling threat displays like gorillas. There are strong matriarchal bonds creating continuous non-sexual associations between mothers and offspring, and weaker sister-brother associations preventing at least some of the potential for incest.

Individual behavior variability, especially learning and reasoning

Primates have considerable ability to vary their behavior individually in response to complex survival demands of the physical and social components of their ecosystem, but they also show the consequences of being the recipients of the personalized, individually directed behavior of other group members.

They may have the ability to learn to use physical structures. Thus many years ago in the laboratory chimpanzees showed that they could fit sticks together or assemble boxes in order to reach distant bananas. Wild chimpanzees use grass stalks to fish termites out of their nests. The skill varies with the individuals. Orphaned young, or young with unskilled mothers, rarely learn the procedure well unless they have an associating kin member who is good at such termite fishing.

In early experiments on the shaping of behavior in primates Harlow and Harlow (1962) demonstrated that infant-mother bonding and social learning were dependent on touch stimulii as well as food supply. The Harlows raised infant monkeys on a variety of mother models consisting of wire screening with and without terry towel wrapping. Infants deprived of the cuddly mother model were fearful and learned little through playful curiosity as they grew. As adults they showed serious mental disturbances in the form of withdrawal from conspecifics, or erratic aggression. Adults who either had a cuddly mother model or the company of peers played and explored more normally. They learned social behavior, including sexual behavior, and were able to maintain normal social responses. The complex of personalized behaviors utilized by different individuals is largely shaped during the intensive learning period of infancy and youth.

Young monkeys and apes learn an enormous amount as they grow. They spend a great deal of time at play. They practice and become skillful at many actions which range from simple running and climbing to recognizing peers, adults and ranks of individuals within their groups. The playing stage, inhibited and harmless though it may be, nevertheless can have substantial effects on the ultimate ranking of individuals in those species that are so organized. Juveniles may be supported by their mothers in play fights, with the result that adult ranks are correlated with mothers' ranks.

It is known that male baboons and macaques often leave their natal groups and spend considerable periods wandering. In the literature there are reports of strangers intruding into baboon groups. Such strangers may achieve high rank quickly (Goodall 1979) or slowly (Strum 1975). Unfortunately, there is only speculation concerning why they were travelling in the first place, and had not remained with their natal groups. We do not know if they were competent low rankers, suffering from alliances. Macaques, in contrast, often wander away from their groups temporarily, and on their return may or may not be able to take up whatever ranks they held before.

One chimpanzee is known to have risen in rank as a result of learned behavior. Goodall (1971) described how a low ranking male, Mike, accidentally clanged kerosene cans together during a threat display and noticed a substantially different effect on the behavior of other group members than when he displayed with almost noiseless branches. He quickly realized the ranking value of the clatter he made with the cans, and by using them in displays he soon became accepted as the group dominant, displacing Goliath who had held the role. Goodall describes Mike as actively seeking the cans prior to displaying. His dominance was eventually terminated even more quickly when, after several years, a young male (Humphrey) charged him, calling his bluff. He fled, unable to back up his displays by physical prowess. In sociobiological terms he was a Cheat, but had acquired his strategy by individual experience, not by inheritance.

With chimpanzees there is some argument over the rewards of high rank. There is enough to eat, and enough breeding opportunity (the genetically determined strategy is for males to grow enormous testicles, thus increasing their sperm count and the likelihood of fertilizing a female during her repeat matings with many males). The rewards may be the attention of others and more frequent grooming. The latter would provide the physical reward of the removal of skin disturbances, but there are many photographs available which show that the rewards of grooming may also include the pleasures of touch and friendship. Rewards for primates appear to include emotional satisfactions as well as material goods such as food.

The process of learning and its association with rewards has been shown among macaques as well as among chimpanzees. Among a band of macaques feeding on a variety of foods at a Japanese beach site an 18 month old female, Imo, learned to wash sweet potatoes and other foods, thereby removing the sand grains. Presumably macaques dislike gritty food, since others of the band would tediously remove much of the sand. Not only did Imo learn to wash her food in seawater, she appeared to like the taste of salt. Two years later she learned to winnow rice grains by allowing them to float in seawater, so that the sand dropped away and the rice could be scooped up. This sequence had some very interesting consequences, which are as significant in understanding the potential brain capability of primates as the original learning.

Imo's mother and sister learned a Cheat strategy. They waited downstream from rice washers and stole the products of the others' labor. Many of the band learned to wash their foods, but the learning proceeded largely down the ranks, not up them. No male older than Imo learned the washing strategy and only one older female, her mother, did. Her juvenile peers saw her and copied, and the skills were passed by imitation through peer play groups. Only when eventual high rankers, who had learned the procedures as young, reached their adult ranks did the cultural practice become established throughout the entire group. This took about ten years. Imo learned a new primary pattern of behavior, others copied so that a group-specific culture was established, and a few individuals learned a secondary skill, cheating. Group cultural patterns can be adopted in some primate species.

A number of chimpanzees and a few gorillas and orangutans have been taught to communicate with trainers. The trainers have often been human families who have brought up a young ape like a human baby although they also gave professional experimental attention to the results. The end result has been the conclusion that there are physical limits on the ability of apes to talk to humans. Primates simply do not develop a voice box capable of making more than the simplest sounds, regardless of whether or not there is the brain power to allow a voice system to realize its potential of language. Other trainers have avoided the anatomical constraint by teaching sign language. Two chimpanzees, Washoe and Nim Chimpsky, developed a substantial repertoire of signs and appeared to be able to invent untaught combinations. They certainly developed very great sensitivity to the intent of their trainers, and could communicate with them in many emotional ways with generalized body language showing excitement, resentment, aggression, dominance, refusal, acquisitiveness, anger, etc. To what extent their undoubted ability to use sign language reflects the ability to formulate abstract concepts, as in spoken language, is controversial.

This short introduction to the wealth of literature on the learning capability and individual behavior variability of primates, especially the apes, and their behavioral consequences, shows forerunners of many of the abilities that we see developed to a greater extent in humans. We need to appreciate how adaptive learning and reasoning can be to animals, especially to the great apes.

Other behaviors

There are a few behaviors which are widespread among the primates, and their presence or absence in humans may reflect adaptations during human evolution. We need to take note of such common primate patterns and why some remain in humans while others disappeared or changed in novel ways.

Most primates are herbivores although some may take grubs and other insects when they are available. Ground dwellers, in particular, will inevitably pick up insects etc. with plants, and chimpanzees may actively seek them. Chimpanzees and baboons, both ground living, are the species known on occasion to hunt and eat the flesh of mammals. The former have a pronounced social ritual of sharing such hunted food. In contrast humans are omnivorous, cultural practices and local habitat determining the foods eaten.

Most primates have strongly developed morphological color patterns of the type which serve as signal badges continuously giving notice of some property such as age, or which are flashed in displays at breeding time or during territoriality. Even gorillas carry such badges, in their case the male age/maturity badge of the silver back. Monkeys have an enormous variety of such badges and a corresponding variety of stereotyped courtship and territorial displays. In contrast humans have few such features which might function as signal badges. Grey hair, eyebrows and (in males) baldness are possibly such badges.

The primate line of descent

The primate adaptive phylogeny is modelled in Figure 11.4.

Initially in the Cretaceous, about 100 Myrs ago, the stock was a form of the small, secretive, generalized mammals, which were capable of moving over ground and other firm structures such as the bases and trunks of trees. Some of these forms developed the ability and behaviors to stay more or less permanently off ground on the vertical structures or in the associated branch network. Then came other adaptations which increased their capability to make their living on smaller and smaller branches of the canopy network. It was at this stage of adaptation to the canopy that the primate characters evolved. These were the characters facilitating three-dimensional locomotion, body flexibility, stimulus sensitivity and social organization. The adaptations included a substantial rearrangement of the body parts, freeing the torso from its horizontal position by making adjustments to the central skeleton and limbs in terms of the size, shape, articulation and form of extremities. The prosimians show the sequence from branch running insectivore - like tree shrews - to canopy inhabiting monkey-like forms, lemurs and tarsiers. A few of the prosimians showed another common continuing trend within animal lines of descent, that of growing large. Eventually they were restrained from using the canopy because of their weight, and had to adapt to specialized trunk living or reverted to the ground.

Meanwhile, within the physically and biologically complex habitat of the tropical forest canopy, sense organs were

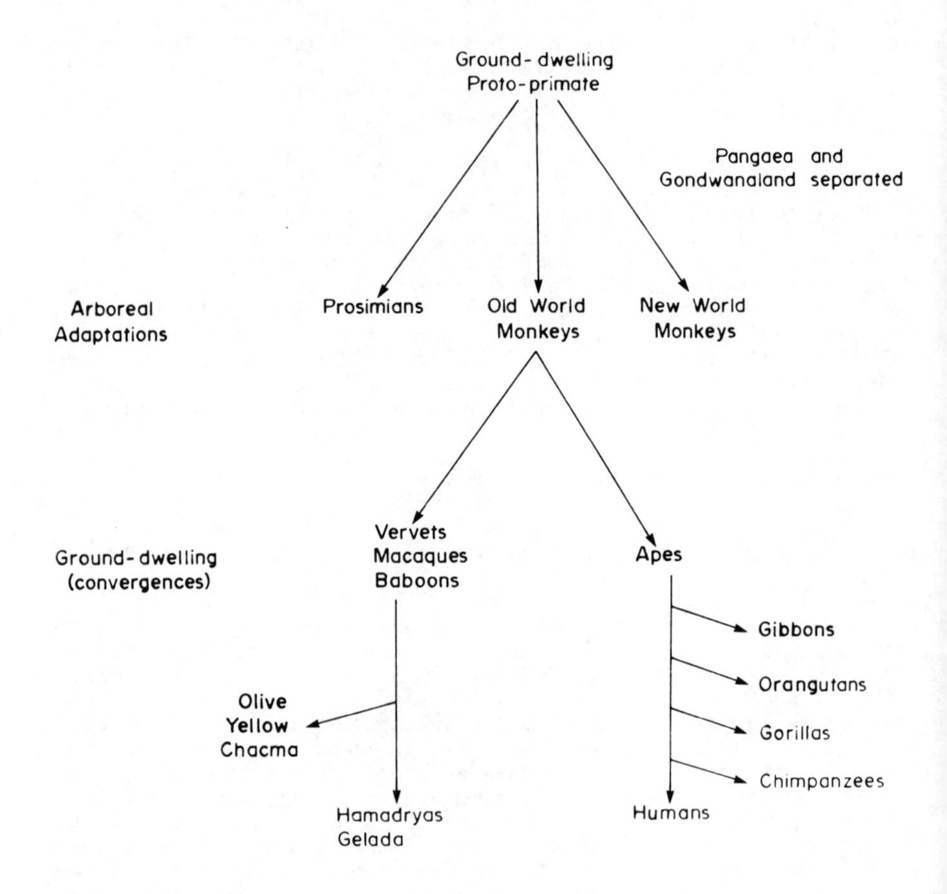

Figure 11.4. Convergences from the primate stock.

refined to three-dimensional perception and the relatively rounded primate head appeared with its directed eyes, backed by a large brain capable of storing information for varying individual behavior. The habitat provided a high diversity of plant and insect foods, important vegetable food being scattered in multiple species stands. There were many poisonous food forms, and cryptically colored but less agile ambush predators such as snakes. The food was more available during daylight hours than at night, and color vision was an important sensitivity. Social living, as with many active day animals, provided important protection against visually detectable predators, and aided in food seeking. Probably many forms developed early, with group size and type of organization varying as adaptations to particular canopy habits. The drab or cryptic coloring of the secretive insectivore ancestor was replaced by the species-specific badges which are now so prevalent among the monkeys and prosimians. Vocalizations developed partly as anti-predator warnings and extended to social calling which assembled group members when food was found. Finally, calling could become intergroup as well as inter-individual as groups split and maintained contact as neighbors. A sensitivity to individual recognition developed based on fine details of facial structures and posture so that mates, infants, rankers and strangers were recognized and a complex social life based on small intimate groups developed. Langurs even distinguish some neighboring groups as enemies sometimes, but not always, open to attack.

With this range of adaptations the stock was evolving towards effective canopy living during the Cretaceous/Paleocene, when the physical separation of Gondwanaland finally split the stock into three. The small Madagascar drift sector broke away with only the early prosimian forms and these, lacking competition from the later evolved fitter monkeys, survived and radiated there in ways no longer seen in Africa.

The western drift sector took the primate stock which became the New World monkeys with surprisingly restricted descent and subsequent distribution. They evolved the prehensile tail but no large ground dwelling forms able to extend the range and distribution of the stock as in the Old World.

The central sector of Gondwanaland retained a monkey stock which competed well within the canopy with other non-primate forms such as the squirrels, flying foxes, birds and butterflies. They diverged into various specializations in canopy living, leaping, night living and group living. Land connections allowed dispersal to forests of Eurasia. Two branches of large forms appeared and converged on ground dwelling.

One of these branches is now represented by the series vervets, macaques and baboons. The macaques in particular found a wide geographical range for their combination of individual adaptability and social organization. Baboons remained within the ancestral African range for some reason, but evolved species completely divorced from arboreal living. These species have the unique social behavior of alliances of friendly males, and frequent overt aggression following threats to enemies, subordinates and strangers. The ability to recognize individuals had extended to friends as well as mates, rankers and strangers.

A second branch of the Old World stock (the apes) also grew large and diverged. The gibbons and siamangs modified their structures to the special requirements of brachiation so that they could swing rapidly through the finer strands of the canopy network between the work stations of firmer or stronger structures. The orangutan is slow and cautious and can test how far off the main strands into the canopy web it can penetrate. The two types, somewhat surprisingly, are either solitaries or small group species within the canopy of dense tall forest. The gorilla and chimpanzee converged on the ground dwelling strategy, but have a social order very different from that of baboons and macaques. Nevertheless, chimpanzee social life is as complex as that of baboons. It is based on flexible group size and the individual recognition of kin, friends, rankers and group enemies. The gorilla, being relatively slow and unaggressive, appears to have a fairly simple social life of one male-several females organized with mild dominance. Both types have progressed only part way to the ground dwelling strategy and live in forested areas, spending some time in the canopy, at least as far as they can extend with their weight.

This Hominoid taxon includes humans in view of our structural similarity, but the specifics of evolution from the Hominoid stock to ourselves is the subject of Chapter 12.

In general though, we must consider the extent to which structures and behavioral qualities allowing the convergence of large primates on ground dwelling were already present in the monkey stock and to what extent the stock was pre-adapted to the demands of ground dwelling when large forms evolved. The new ground dwellers were in competition with efficient herbivores and omnivores already there, and were susceptible to the many predators living on the bounty of African ungulates. Large monkey sized prey moving in slow groups would seem to be easy pickings for rundown and ambush type cat, dog and other predators.

It is important to remember that there are several distinct types of ground habitat such as parkland (trees and grassland interspersed), savannah (grassland), wetland, beaches, scrub and desert. The habitat may be extensive or patchy with many ecotones. There are some adaptations which would facilitate life in any of these conditions. They are an alertness to predators at a distance, group living, learning and intelligence sufficient for knowledge of habitat divisions such as food sources and seasonally varying home ranging, ambush opportunities for self and predators, and hiding and sleeping places. Adaptations also include long and short-distance acute perceptibility, particularly with sound and vision, and flexible circadian rhythms. Almost all of these characters are shown by baboons and macaques, and chimpanzees and gorillas. The exception appears to be the flexible circadian rhythm. If the circadian rhythm is inflexible it could constrain feeding and mating resource opportunities largely to daylight hours. Flexible behavioral rhythms, allowing a ground dwelling ape to vary its activities over 24 hours, would permit it to be active at dawn, during moonlit nights, or for irregularly appearing swarms or flocks of prey species. Sleeping during the dark hours as a means of conserving energy when unable to feed (due to undetectable nocturnal predators) limits the activities of hamadryas baboons and geladas to within daily range of safe roosting cliffs.

In conclusion, much of the anatomy and structural arrangement which could facilitate reentry of the primate stock to a ground dwelling but novel way of life was a product of arboreal adaptation. Once large primates appeared, there was not only the opportunity for them to revert to ground dwelling but also the need, since they could only be active arboreally in the largest trees, which in turn were a product of a specialized high and dense forest habitat, and not continuously present in parkland and bush country. The monkey stocks which went through the enlargement process in the canopies of a low sparse forest would be forced to adapt to trunk dependency or ground utilization. This process has been followed at least three times with varying degrees of success: the Indri (a prosimian), the vervet/macaque/baboon descent, and the gorilla/chimpanzee/human descent.

Summary

The monkeys and apes are tree dwellers, with adaptations which turned out to be pre-adaptations for ground dwelling when large forms evolved and were restricted in their use of the forest canopy. The pre-adaptations included body flexibility for three-dimensional living, including upright extension, and appendage flexibility with grasping digits. They also evolved outstanding visual sensitivity, utilizing many of the perception options of light, e.g. color and acute focussing vision. This required substantial brain capacity, which in turn allowed the processing of other complex information which permitted individually variable behavior through memory and association, and the development of complex social organization based on the recognition of individuals and subtle body language displays rather than overt threats.

There have been at least three adaptive lines to large size, two of which have converged on ground dwelling: the prosimian Indri, the Old World vervets/macaques/baboons, and the ape line of descent gorilla/chimpanzee/human. The New World division of the primates did not produce a ground dwelling line, and their distribution remains limited to forest canopies.

CHAPTER 12. HUMAN ETHOLOGY.

A part of understanding human behavior is considering whether at least parts of it have descended from the adaptive behaviors of ancestors in pre-human and prehistoric ecosystems, and if so how the behaviors have diverged and radiated to their present complexity in different cultures.

There may have been many popular attempts to describe human behavioral evolution. Biologically informed popular books started with *African Genesis* by playwright Robert Ardrey (1961) and were followed by his series *The Territorial Imperative* (1966), *The Social Contract* (1970) and *The Hunting Hypothesis* (1976). In the meantime mammalogist Desmond Morris was writing the series comprising *The Naked Ape* (1967), *The Human Zoo* (1969), and *Intimate Behavior* (1971), and continues with illustrated books such as *Manwatching* (1977) documented by candid photography. Professional anthropologists and biologists have provided more or less technical and highly readable versions, such as Lionel Tiger's *Men in Groups* (1971), Wilson's *On Human Nature* (1978), and the Nobel Prize winner Lorenz's *On Aggression* (1966). They are still being written: *The Red Lamp of Incest* (Fox 1980), *The Evolution of Love* (Mellen 1981), *The Sex Contract* (Fisher 1982), *The Monkey Puzzle* (Gribbin and Cherfas 1982) and *Subordinate Sex* (Ellis 1982). Some versions are even rated by those who should know as having literary merit, e.g. William Golding's *Lord of the Flies* (1954) and *The Inheritors* (1955).

They are fun to read. But the reader should take care to enjoy him or herself even more by critically appraising what is written with what can be deduced about the behavior of survival machines in the near-ape form, the pre-human form, and the human form. For the human form we must consider behavior in very early stages of environmental manipulation, when our ancestors were equipped with only a few tools, weapons, energy sources (fire), early domesticated animals, simple home-habitats and their population density was low. The ecosystem we now inhabit is a product of several thousand years of environmental impact. This is only a few hundred generations of human evolutionary time. It is believed to be too short for significant adaptive change by ourselves to the oddities of our rapidly changing ecosystem.

The popular books can be criticized, or feared, as a danger to society. It is in the area of behavioral speculation that biological support can be theorized for insidious human social prejudices such as racism and sexism. Even worse, the theories can supposedly support group pathologies of torture, murder, rape, war and genocide. The social Darwinism of 19th century industrialists, which remains with us today, is relatively harmless compared to the genocide clearly proclaimed by Hitler (1939) in *Mein Kampf*, but largely ignored until ferociously implemented during World War II. It is important that we critically appraise even jolly attempts to speculate about the origins and inevitability, the biological determinism, the flexibility or inflexibility, of human behavior.

The need to be critical, so that we are not misled, becomes even more important as professionals expand from writing books and lend their knowledge to producing cinema and television spectaculars. The days of the obviously nonsensical *1,000,000 Years B.C.* and *Planet of the Apes* have been superseded by such films as the realistically intended *Quest for Fire* (Rosny-Aine 1982). Even this film, supported as it is by consultant behaviorist Desmond Morris, and linguist Anthony Burgess, is only partly the product of the knowledgeable working to the constraints of the film medium.

These books and the films can be judged against the criterion of the ethogram. How could evolving ape-humans have survived at each distinctive stage as individual animals with a set of behaviors successfully balancing the diverse demands of the ecosystem in which they were living? At what stage in human evolution did our progressive environmental impact from manipulating habitat and organism affect the process of natural selection? How did our own evolution contribute to diversifying behaviors between evolving cultures so that any inherited originally adaptive components of behavior held in common by humans became overlaid by innumerable culturally and individually acquired expressions?

At some stage in the evolutionary series, the animal behaviorist must accept that the sociologist, anthropologist, psychologist, archeologist, economist and historian have relevant detailed information for understanding the processes of cultural change and survival pressures on humans and their groups. The behaviorist must decide once again where he personally wants to limit his understanding of the processes by assigning them to black boxes within a model of human evolution.

To make this decision, the behaviorist needs to have some knowledge of the fossil stages connecting the primate stock with humans, the ecosystems in which the ape-humans lived, and the types of behavior which they must have shown. The ethogram concept can help us with the latter once we can deduce from fossils, artifacts and stratigraphy information about geological time and climate, food and habitat, tools and weapons, environmental impact and social organization.

The fossil record

Figure 11.1 provides a summary of the fossil record of the primates, including the line leading towards apes and humans.

An important pattern in the early record of the Hominoid line is the diversity of forms in the Oligocene and Miocene, from about 36 Myrs to 13 Myrs years ago, followed by the virtual absence of fossils in the Pliocene (Pilbeam 1984). Yet since the Pliocene, through the 3 million years of the Pleistocene, we have the great apes, modern man and a number of fossil forms of both. What happened in the Pliocene is one of the key questions in human evolution.

Prior to the Pliocene, and unlike today, there was a variety of apes. These Pithecine fossils are grouped in several genera such as *Dryopithecus*, *Ramapithecus*, *Aegyptopithecus*, and *Proconsul*. These were tree dwelling, moderately large ape-like monkeys or monkey-like apes which probably used the ground considerably and lived in forms of social groups similar to today's chimpanzees.

At the end of the Miocene period, about 17 Myrs ago, as Africa and Arabia drifted into contact with Asia, the early apes were introduced into a new kind of forest. This was country which was affected by annual seasons, hence the woodlands could be extensively deciduous, with substantial changes from season to season in the food available to tree and ground dwellers. The Siwalik fossil beds of Pakistan have shown that members of the genus *Ramapithecus*, towards the end of the Miocene, had robust worn teeth, suggesting a diet shift from soft plant food such as fruit and buds to harder products such as seeds and tubers.

In Africa the Pliocene was a hot dry period during which the continent became extensive desert and semi-desert. The forest ecosystem was reduced to restricted heights or coastal regions where combinations of humidity or temperature could continue to provide adequate moisture and other environmental conditions for tree growth. There have been no fossils found until very recently.

There are now reports of a few fossils, probably on the human line of descent, from the Samburu hills of Kenya, dated about 8,000,000 years ago (see Figure 12.1 for a map),

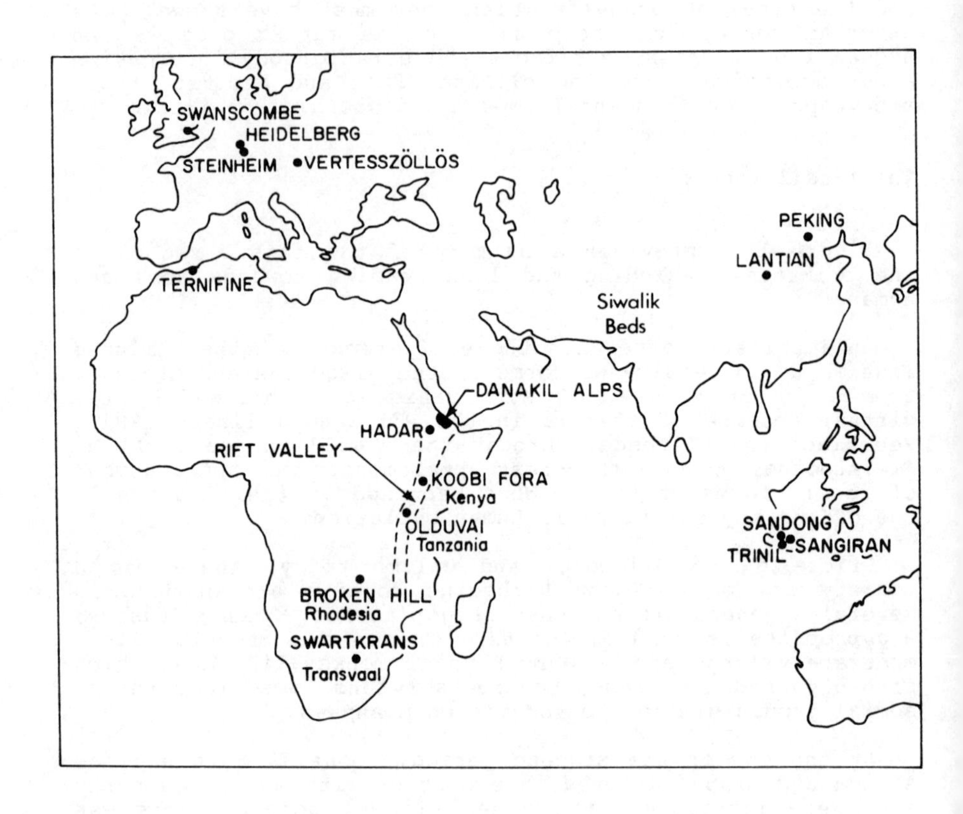

Figure 12.1. Map of the distribution of ape and human fossils.

but little is known about them at the moment. We do not know if the lack arises from poor fossilizing conditions, or if archeologists are digging in the wrong places. Deducing where the fossils might be and obtaining funding and permits for exploration is part of the excitement in this subject.

As the super-tropics of the Pliocene in Africa cooled down and gave way to the colder climate of the Pleistocene and Recent eras, fossils reappear in the scientific record. They all show clear pelvic evidence of two-legged locomotion, and hence are on the human, not the ape, line of descent. Gorillas and chimpanzees do not fossilize well. The groups which appeared are shown in Figure 12.2 at 3-4

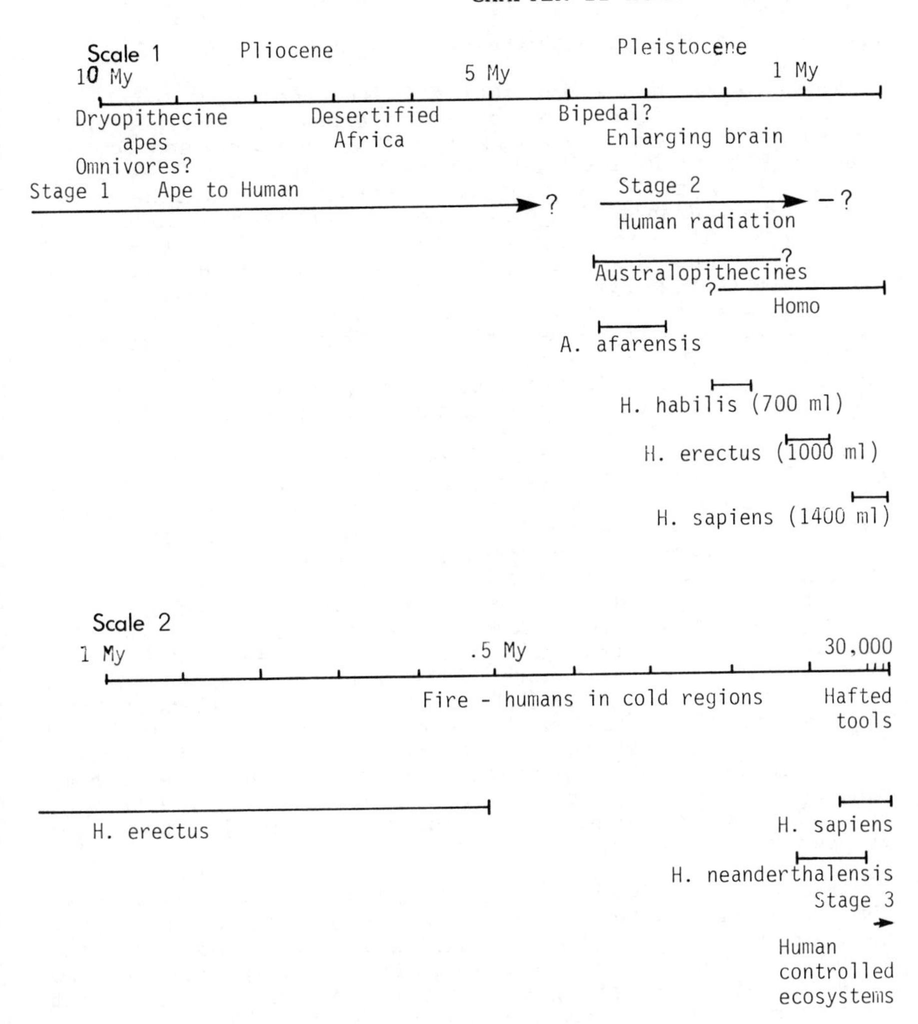

Figure 12.2. The human fossils and summary of evolution.

Myrs.

There are two main forms, which for simplicity we will refer to as the two genera *Australopithecus* and *Homo*. The specialists can get into complex arguments over whether the differences between the two lines are equivalent to generic differences between extant primates or even arguments about the validity of species within each genus.

There is an intriguing set of fossils from the Hadar area of Ethiopia referred to as *A. afarensis*, and popularly called Lucy and the First Family. They are supposed to have died together (Johanson et al. 1981). At about 3 Myrs ago or more they could have been ancestral to both lines. At Laetoli in Tanzania there is a set of three human footprints 3.5 Myrs old (Hay and Leakey, 1982). These look suspiciously as though they were made by a large father and smaller mother walking side by side, with a playing child stepping carefully in the father's prints after him. The males of *A. afarensis* were 50-100% larger than the females, suggesting that their social organization was polygamous and based on power and displays as in chimpanzees and gorillas. They appear to have lived in grassy woodlands.

Enough Australopithecines referred to the period 3-2 Myrs ago have been found in East and South Africa that they are often grouped into two species, *A. africanus* (the gracile form), and *A. robustus* (the robust form). The latter may include the other named species *A. boisei* and *A. habilis*, but *habilis* may be included in the genus *Homo* and gives rise to arguments whether *Homo* and *Australopithecus* were contemporaries or not. *Homo* (or *Australopithecus*) *habilis* may have been a carnivore according to its teeth, which show evidence of grawing on bone. The brain was larger than in other Australopithecines (about 700 cc, compared to 500 cc in the other species) and there are associated used and altered stones brought from distant sources. *H. habilis* could have been carrying objects around, which suggests home ranging from a base, with hunting and gathering, provisioning of mothers and infants by bonded males or grouped kin, or both.

Meanwhile, according to the limited fossils available, *Homo* took over as the predominant fossil in beds more recent than about 1 Myrs ago. *Australopithecus* died out, for unknown reasons. The first well described species of *Homo* is *H. erectus*, a large, undoubtedly tool using human with fossils distributed from Java to Israel and Africa. It had a brain volume not as great as that of modern man, at 1,000 cm^3 compared to ours of 1400 cm^3.

Homo erectus includes the Peking man named by Davidson Black, a Canadian medical professor in China. Although the actual fossils went astray during the Japanese occupation of China during World War II, many casts of them were taken and the cave site provided cultural evidence of fire, and possibly ritual cannibalism (skulls broken open from the base). The sites are now being intensively investigated by Chinese scientists (Wu et al. 1983). *H. erectus* lived in temperate climates throughout Africa and Asia but did not succeed in dispersing itself to the American continent and

Oceania. That range extension required further adaptations achieved by yet another human species, *Homo sapiens*.

Within the last hundred thousand years, two closely related forms with similar brain capacities appear in the records largely in well searched Europe. These are *H. sapiens*, modern man, and *H. neanderthalensis*, both with considerable ability to fashion tools and weapons, and even create cave paintings and bury their dead. There is much speculation about the intelligence, sociology, cultures and ecology of these prehistoric humans. From the animal behaviorists' point of view, important deductions are (1) the extent to which these people lived in separate kin groups and how much exchange of genes and cultures there was between the groups, and (2) was the social structure essentially monogamous (as now) with two-parent raising of paternally accepted children. Answers to (1) lie in the extent to which artifacts and artwork of various groups were similar, and one answer to (2) in the relative stature of males and females. Similar artifacts and artwork over great distances, would suggest considerable biological and hence cultural exchange between groups, and similar sized males and females as today would suggest primarily monogamy, not polygamy. The two species appear to have coexisted, but *neanderthalensis* disappeared for unknown reasons.

The archeological record of domesticated species and complex artifacts from approximately the past 10,000 years since the last glacial period, demonstrates yet another phenomenon of human descent. Humans have substantially controlled and managed (and in recent years often mismanaged) their ecosystem. Over this period of time, humans have virtually eliminated predation on themselves, minimized exposure to killing weather and pathogens, raised primary and secondary biological productivity ensuring locally abundant food supplies, and developed the capability for global emigration and population dispersal. As a result the process of natural selection on humans must have undergone substantial changes in ways not clearly understood. If parts of our culturally shaped behavior were once adaptive it does not follow that they now are. Ten thousand years of a rapidly changing ecosystem and protection against natural selection involves about 500 generations, considerably more than needed to select for 100 years of industrial melanism in the peppered moth and other insects (see Chapter 10).

In summary, of the fossil and archeological evidence, there appear to have been three distinct periods of adaptive change in the human line of descent. These ar:e (1) change from ape to human, i.e. from forest to woodland niches, (2) an early diversification of human lines into several niches, including at least one line of meat eaters with extensive dispersal into new areas and (3) survival within a progressively human controlled ecosystem.

Finally, it should not be overlooked that apart from the half skeleton of Lucy, most hominid fossils are fragments of disconnected teeth, skulls and limbs, a poor basis for reconstructing whole animals and their behavior. That is why our ideas about our prehistory must change with new discoveries.

Subhuman to human behavior

Most of the popular authors have concentrated on the generalized change from arboreal to ground dwelling primate, rather than the change from hunting-gathering pre-human to hunting-gathering human, which does not involve an ecosystem shift. Only a few authors have attempted to place where the change occurred rather than in what general kind of habitat. Also, not many of the popular authors have considered the lives and adaptations of group members other than the adult male.

Orthodox and unorthodox theories

The deduced generalized line of descent, with associated behavior, which has been accepted by so many authors without considering a long-standing but viable alternative, can be called the Orthodox theory.

This orthodoxy claims that an occasionally ground dwelling early ape, which was rather like a chimpanzee in habits, progressively spent more time on the ground in fringe areas of the forest. It divorced itself from forest canopy and undergrowth plant food and converted first to meadow plants and roots, added insects, birds' eggs, small mammals and other creatures, and then scavenged meat from large carcasses left by predators. Gradually increasing skills of cooperation and the use of stones and branches as weapons helped the ape-human to drive bigger and bigger scavengers away from carcasses, and allowed him to ambush in groups and run down larger and larger prey. The reward for cooperation between socially bonded male hunters was access to food and sex, the latter provided by a monogamous social suborganization within male ranking groups. The females adapted by providing unlimited sexual rewards through eliminating an estrus period. On the way, the ape-human lost its hair as an adaptation to not overheating while on the hunt.

In recent years there has been some acknowledgement that females comprise about 50% of the species, and that a majority of individuals are not adult male hunters. There has been some recognition that each individual at all life stages will probably have evolved specific adaptations favoring their survival. Infants, juveniles and the elderly (experienced), not just the adult female mother, live in different worlds and different sub-ecosystems than that of the adult male and we can expect them to be adapted to their sub-ecosystems. As a result, the orthodoxy has recently come to admit that females probably contributed greatly to the food supply by socially organized gathering, bringing in the daily rations of roots, fruits, grubs and small mammals. Not only that, but carrying devices were probably invented (by women, or even children) before or simultaneously with tools and weapons. There are many obvious sophistications to be added to the orthodox theory. These involve crediting the surviving elders as rare and respected stores of experience and wisdom, and the newborn infant as a survival machine looking after itself by evolving grasping power which allowed it to hang on to branches, and mother's hair and breasts while she moved around.

The unorthodox theory postulates a non-fossilizing coastal amphibious pre-human during the later part of the Pliocene period from 13-3 Myrs ago. As the heat of the Pliocene reduced the extent of African tropical forests and converted the continent largely to arid desert or semi-desert, so those ground dwelling primates which happened to inhabit the coastal fringe area and were already utilizing the beach ecosystem as an occasional source of food, were able to maintain themselves there in spite of the heat and loss of terrestrial food supply. They became separated from the montane forest enclaves far inland, and could evolve their own adaptations which were different from those of forest apes.

The original proponent (Hardy 1960) pointed out that an amphibious adaptive period would explain the human insulation of subcutaneous fat rather than fur, the streamlined arrangement of body hairs, the diving reflex controlling breathing and vascular constriction and a number of other characters inexplicable as adaptations to continuous inland ground living. The reviver of the unorthodoxy (Morgan 1972, 1982 and 1984) extends the theory to point out how such a period could account more satisfactorily for different hair patterning in male and female (floating babies clung to long head hair), ventral-ventral mating (a vaginal shift from sitting hairless on sand), the loss of olfaction, refinement of calls into language, and lack of a salt satiation response.

The difficulties with the theory need to be appraised to see if they are insoluble, or as simpleminded as some of them seem. Just where in Africa was this select coastal stock? (The same difficulty applies to the orthodoxy. Just where were the ground dwelling apes, and all the other primates for that matter, during the forest decimating heat of the Pliocene?) The most serious difficulty with the unorthodoxy arises with the coastal ecosystem being in a tropical region. Hot coastal ecosystems of surf beaches, lagoons, estuaries and wetland are loaded with predators and other environmental differences from forest, parkland and grassland. Surf beaches are very noisy and the swash zone is physically dangerous, yet they provide rich food supplies, especially at low tide. Lagoons and estuaries are quiet so that vocalizations would be heard easily enough and would be needed for a semi-submerged amphibious ape. But they are very dangerous due to sharks, crocodiles, hippopatamuses, and other carnivorous or large animals. Wetlands and marshes (fresh or salt) might meet many of the requirements for a relatively safe and food bearing habitat, with high nocturnal activity of potential prey, from which groups of small irregularly active vervet sized primates (with a high birthrate counteracting high mortalities) could range out to nearby surf beaches and lagoons during daylight if they so wished.

The essence of this argument is that both orthodoxy (the Tarzan the Mighty Hunter theory, so named by Morgan 1972) and unorthodoxy (the Aquatic Ape theory) and any development or alternatives need critical appraisal. The appraisal should be made in terms of the total ecosystem and the total behavior of the individual animal attempting to stay alive at all life stages and ensure perpetuation of its genes in another generation. Any reconstructed fossil or hypothetical ancestor needs to be considered as a living individual capable of a balanced set of adaptations to complex environmental properties.

The type of detail which must be considered in terms of either orthodoxy or unorthodoxy are such matters as the extraordinary genetic similarity between chimpanzee, gorilla and human (Griffin and Chervas 1982). They share 99% of their DNA. Accordingly the species' differences need to be considered in terms of regulating genes affecting relative proportions of structures by allometric growth with size. Or neotenic changes in relative time of appearance of structures during individual development might have occurred. In all cases, though, the individual animal must remain viable, even though this might be brought about by an increased dependent infant stage. Gestation could not increase indefinitely, since the fetus and its increasingly large cranium must pass through the female pelvic canal at birth. Also, certain cultural developments could not occur until certain body sizes, and possibly brain sizes, had been reached. Thus effective hunting spears require a body size capable of generating leverage enough for a relative heavy weapon. There are many such important details.

Details of sexual organs and behavior, and phylogenetic reasoning based on them, have already been given in Chapter 10. The argument can be made objectively that the social organization of the common chimpanzee-human ancestor, was based on overtly ranking behavior between males. Male-female size differences in *Ramapithecus* support the phylogenetic reasoning. The unique (apomorphic) sexual and other behavioral characters of humans developed since the common ancestral stage with chimpanzees must be explainable in terms of a sequence of viable pre-humans and humans in describable habitats adapted to comprehensible ecological niches.

Locations

Kortlandt (1972) is one author who clearly states where the Pliocene ground dwelling habitat shift occurred. He places it in the central East African region, extending from what is now Tanzania to Ethiopia, where the high irregular country of the African rift valley should have supported relict montane cloud forests. These stocks would allow monkeys to survive in the canopy, and Hominoids first in forest clearings and parkland, and then on the open grassland. The present generation of fossil seekers are in the right region.

Morgan (1972) did not specify where the aquatic ape may have spent its coastal period. We badly need some knowledge of African coastal geography during Pliocene times to appraise whether appropriate habitat actually occurred and where. It seems conceivable that the aquatic ape could have survived on the whole coastline of South and East Africa, or even on the neighboring coast of the Arabian Sea. Even in the driest countries, without rivers and estuaries, lagoons and salt marshes occur. The shoreline stocks in the increasingly hot Pliocene would eventually have been separated from the montane forest ancestor, allowing conventional speciation to occur. However La Lumiere (1981), republished in Morgan (1982), hypothesizes an isolated region: the Danakil Alps near the mouth of the Red Sea. These mountains would have been an island or islands during Pliocene sea level changes.

One of the disadvantages of unorthodoxy, even when it expands on the ideas of a respectable marine biologist (Sir Alastair Hardy 1960), is that nobody takes you seriously enough to tackle the challenge of obtaining funds and archeological permits to look for what is believed to be the unlikely occurence of fossils in old beaches without some guidance concerning where on the continent to look. But someone should be looking in the Danakil Alps.

By the time of *Homo erectus*, 2-0.5 Myrs ago, the hominid stock had spread itself throughout tropical Africa and Asia (Figure 12.1) and into some of the East Indian islands. More recent fossils of Neanderthal and modern man show that humans were spread throughout Europe, Asia and Africa. The spread to America and Australia was in Pleistocene times, in the former via the Bering Strait bridge during glacial period exposure, and in the latter by manned boats.

Vignettes of social behavior

Some scientists have taken to using the literary device of the vignette to present their concepts of the social behavior of evolving ape-humans. Fisher (1982) provides them for *Dryopithecus*, unnamed pre-humans of 10, 9 and 8 Myrs ago: *Australopithecus*, and *Homo erectus*. It is a technique which concisely describes important biological functions with subtle implications and atmosphere. The technique provides a way of condensing the professional's ethogram into popular format. It needs expansion which, in Fisher's case, she assists by providing a table of time, environment and traits. The vignettes are very short but should be read with care, as any fiction should be. The point by Fisher is that they present the largely ignored adaptations of female and young rather than those of adult males.

Homologies or analogies?

There are many similarities in the behaviors of animals and humans. These range from the simplest action of sneezing to the fact that we can apply the concept of courting to both human and animal mating. Some of these similarities have generated a great deal of emotion and argument about their significance at times when anti-social consequences such as racial exploitation had been feared. Some of these topics are presented here to show the nature of the emotion generated, and to indicate where and how the behaviorist may wish to consign his level of understanding of the causes of the human behavior to a black box compartment within a generalized model of human behavior and its evolution.

Homologies

There are undoubtedly behavioral homologies between humans and other mammals and between humans in different cultures, since some behaviors are completely dependent on mammalian anatomy. We eat, drink, defecate, belch, fart, breathe, sneeze and mate using the same body organs as any other mammal.

There are some other behaviors which occur time and time again among mammals due to common structures which have adapted in many different ways. We scratch ourselves to remove an itch, using whatever limbs or movements best serve our needs, and we are not averse to helping a colleague in need. The scratch of chimpanzees and humans is homologous. But at this point in the similarities, we reach the arguable: have we, in many of our cultures, diversified and ritualized the friendly skin care grooming which occurs among apes and probably occurred in our pre-human ancestors, into the professional attention of hairdresser, masseur, physiotherapist and even dentist and dermatologist? Many people enjoy these attentions but others don't. Professional skin care serves the same function as the grooming of chimpanzees. It could be just analogous. But is it homologous in the sense that skin care is an essential behavior, and we seek means to meet this need according to the culture in which we find ourselves? This is the kind of argument underlying the intriguing similarities described by the popular authors.

The argument that the ecosystem places demands on survival machines which are met by humans in ways which are homologous with those of other mammals (especially primates) has objective support in at least some simple behaviors involving more than just the physical feeding, grooming and breeding already listed.

Eibl-Eibesfeldt (1970) has documented the global distribution of the eyebrow flash, a greeting display implying goodwill and attention. When people of many different cultures meet or are conversing in a friendly fashion they frequently flash their eyebrows quickly up and down. We normally do not notice the behavior, but it has the characteristics of an animal display. It is a clear signal, in that alternative actions for the same sets of brow muscles include the frown, which has a very different meaning (at least in some cultures).

Morris (1977) documents many human signals which are recognized by people of different cultures, and has even provided a distributional analysis of gesture patterns in Europe (Morris et al. 1981). Human signals which are widely recognized around the world include the welcome (raised arm), the triumph ceremony (both arms raised and a grin), and threats (for example, tense raised arm with digital clench or point). Morris even attempted in the cine-film *Quest for Fire*, to deduce primitive signals which were the basis for later cultural radiation. He had his heroine, for instance, who was lost from her tribe but accidentally assisted by the three males of another tribe, attempt to join their group by alternately following and hiding, and by using gentle calls and waves of the arm.

We also recognize similar badges in different species. Thus gray hair, a widespread age badge in humans, can be seen in dogs and chimpanzees, especially around the bristles, and is localized in gorillas (the saddle-back) and sea otters (the head). In contrast the other age badge in the male, baldness, may not be global. It may be that it is more frequent in some human stocks than in others. In either case, why should such age badges occur? Do they have a signal function today, and did they in the past? It is enjoyable and fun to theorize about the survival value of human badges and signals. For example, inexperienced humans may be able to locate and follow an experienced human in a panic situation more quickly if he or she wears an age badge than if he or she did not. Such fancies are actually scientific hypotheses. Unfortunately they are virtually untestable in human society.

Nevertheless, some of the theories are of sufficient social concern that they need specific presentation.

Aggression, dominance, war, racism and genocide

There have been continuing arguments that aggression is an inherent property of humans.

The basics of the inherent aggressiveness theory are that in our line of descent we became a carnivorous ape basing our group living on male dominance and territoriality. The male had to be capable of overpowering large prey, defending good habitat, and keeping subordinate males under control in their frequent attempts to overthrow him and reap the rewards of dominance. Females sexually selected males for physical prowess, and had to suffer the personal abuse of breeding with males stronger and more aggressive than themselves. The theory is supposed to still apply since present human cultures and population density are so recent that the old survival behaviors have not yet been selected against. The social dangers of the theory are obvious. We could rationalize almost any anti-social violent behavior ranging from child abuse, wife beating and granny bashing through lynching parties to nuclear blasting another nation, and mass genocide. The ideas unfortunately seem plausible, especially in male dominated societies. The counter arguments (Morgan 1972) inevitably are not strong since they are based on such premises that humans in modern cultures are probably the least aggressive of the primates, and there are not objective measures of our aggressiveness similar to those that can be obtained from animals.

A more moderate expression of the concept is that although learning and culture largely determine the degree of aggression in any society, under emotive conditions of panic, starvation, and mass hysteria, inherent actions might emerge from under the cultural inhibitions overlying them and be expressed.

The difficulties of making the arguments one way or the other are compounded by their involvement with race, and the controversies surrounding racial, anatomical, behavioral and intelligence differences. The anatomical differences between white, black, mongoloid and other humans were reviewed in the 1960s (Coon 1965) and earlier. Their survival value as subspecies' adaptations to environment was also worked out. For example, pale skin favors the production of vitamin D in the weak sunlight of northern regions, black skin eliminates the problems of excessive radiation impact in the tropics, and mongoloid compact body structure reduces the impact of extreme cold in continental winters. It seems very simpleminded, but the full arguments have some validity in showing that humans have subspeciated in typical evolutionary fashion, and that subspecies and other smaller stocks (demes) exist today as races. This does not, of course, mean that races are significantly different in intelligence and other characters presently considered as socially and racially desirable or superior. There is some evidence for different levels of infant placidness in different races, but this sort of subspecies research on humans is affected by the opposition of some people who claim that simply investigating the topic opens the way to racist ideas.

In short, the ideas about power, aggressiveness and inherent differences within human society are so emotive that if animal behaviorists attempt to apply their ecosystem and survival concepts to recent times when present cultures were emerging, they should expect controversy. To minimize unnecessary argument, animal behaviorists venturing into human speculations need to keep informed on topics ranging from pre-human fossils and their deduced behaviors, through evolutionary concepts, to information from sociological, anthropological, economic and psychological investigations of human organization. And so do their critics.

Sociobiology: sexism and cheating

Sociobiology has added several controversial concepts to the problem of social Darwinism. The concepts can facilitate rationalizing cheating and sexist behavior in humans.

A Cheat in sociobiology is just another animal that is doing its best to survive and breed, and has adopted a minority tactic which will be eliminated if its net benefit to the individual is zero. It is reasonable for cheating to be adopted by some proportion of an animal population. Cheating in humans has different effects and can produce costly social disruption. There are very different judgments here taking place under a common name.

The genetic advantages of a species having male and female sexes so that gene flow and modifiability are favored have led to the concept that a two-sexed species can adapt to two different strategies. One sex can invest its energy in producing many small gametes and distributing them widely and the other sex can invest its energy in producing a few large gametes loaded with a food store. The two strategies are incompatible in species where parental behavior by both breeders is important to the survival of the young. The female needs the male to invest its energy in parental care rather than in further gamete distribution, which would take up its time and produce half sibs competing with her brood. In humans, this argument can lead to the rationalizing of male promiscuity, and the inevitability of single parent families of female and young, while males live independently in a society which provides more material rewards to males than females.

It seems that advances in understanding animal behavior will continue to tempt biologists to extrapolate to human social organization, and we should expect continued critical responses from non-biologists.

The value of a human ethogram

It can be revealing to build an ethogram for humans as well as for animals. Any such ethogram will undoubtedly reflect cultural practices, but it will direct attention to behaviors which may be similar in form or function to those of animals, or even more important may show unique distinctions which may have an adaptive explanation.

Humans show, for example, regular 24 hour behavioral cycles of activity and inactivity (sleep, siesta, etc.). The periods may differ in time and duration by culture, but there are some similarities. In general humans are awake during daylight, but instead of sleeping throughout the night like other day-active animals, humans extend activity into the early hours of darkness. And they can be relatively easily reactivated in the dark under special conditions. Is an approximately 16 hours daily activity phase adaptive? Or was it adaptive during the period before we started to control our ecosystem? As with so many other behaviors, it is not difficult to postulate some adaptiveness. Nocturnal activity would open up new sources of food, including the many abundant dusk-active insects and small mammals, and especially during moonlit nights. Ready wakefulness at night would increase survival against nocturnal predators.

Another type of question can be put in a slightly different form. The Aquatic Ape hypothesis explains the evolution of the human nose as a valve assisting in keeping water out of the nostrils. The nose of the swimming proboscis monkey, *Nasalis larvatus*, is convergent, as is shown by its presence in the females and young. However, the males have evolved the nose into a massive secondary male sexual character, which they presumably flash in some form of courtship or threat display. Why does the human male not have a corresponding pendulous nose? Presumably our social organization of relatively peaceful monogamy without continual defense of the female by her bonded male, does not require such a displayable sex badge.

Understanding our own behavior

We should ask the question whether investigating human behavior by the objective procedures applied to animals, and interpreting the results with concepts of adaptiveness, helps us understand our own actions, particularly our far-reaching social and environmentally impacting actions. Do we benefit from a generalized appreciation that human behavior may be adaptive to its culture and either descended from ancestral ecosystem adaptations or convergent with animal responses to similar environmental situations? Do we benefit from predicting our own behavior and that of kin, colleagues or hostiles when in situations where animals would act in statistically predictable ways? For example, is it useful to us to acknowledge that the way in which people rank and defer to or dominate others when in groups, is similar to ranking in animals? If we live in a culture where territories of home, a regular seat on the bus, or national homelands are allowed by custom, and in some cases by law, does it help us to realize that our culture has converged on this territorial pattern because it serves the same function as in animals of more or less peacefully allocating space between competing or cohabiting conspecifics?

The questions may be ignored, but they are there implicitly being asked of each individual biologist. Our answers should be based in part on an informed understanding of the biological facts and concepts, as well as the social and ecosystem consequences of our actions.

Finally, increased technology, both inanimate and animate, will raise more and more ethical questions in human behavior, to answer which the biologist, including the behaviorist, may have to provide some specialist information. The questions include ones like "To what extent do we need to conserve species' gene pools by preventing extinctions or excessive harvesting?" Some of the questions are frightening. For example, "Since genetic similarities between chimpanzee and human are so close, can we and should we hybridize the two?" The idea and consequences were explored in novels long before recent artificial insemination and placental implantation techniques circumvented courtship difficulties and made the question real.

Summary

Human behavior shows similarities with that of animals, but also radiation into culturally diverse forms. Animal and human comparisons have the potential to suggest whether some behaviors are homologous in the sense of being continuously present during descent from a common ancestor. The comparisons may show that some behaviors are analogous in that humans under evolution converged on behaviors which other species found to be functional in similar ecosystem situations. It is important in such comparisons to know as much as possible about the fossil record for the ape-human line of descent, and to develop efficient hypotheses about the social organization and behavior of each recognized fossil form.

Biologists who make such comparisons must be prepared to bring into their data base relevant information on social organization and behavior from anthropology, sociology, and psychology, and guard against their cultural and sexual prejudices.

REFERENCES

Aidley, D.J. (Ed.). 1981. Animal migration. Soc. Exp. Biol. Seminar Series No. 113. Cambridge University Press, Cambridge. 264 pp.

Alkon, D.L. 1983. Learning in a marine snail. Sci. Am. 249(1): 70-84.

Ardrey, R. 1961. African genesis. Atheneum, N.Y. 380 pp.

Ardrey, R. 1966. The territorial imperative. Atheneum, N.Y. 390 pp.

Ardrey, R. 1970. The social contract. Atheneum, N.Y. 405 pp.

Ardrey, R. 1976. The hunting hypothesis. Atheneum, N.Y. 242 pp.

Argue, A.W. 1970. A study of factors affecting exploitation of Pacific salmon in the Canadian Gantlet fishery of Juan de Fuca Strait. M.Sc. thesis, University of British Columbia. 259 pp.

Audley-Charles, M.G., A.M. Hurley and A.G. Smith. 1981. Continental movements in the Mesozoic and Cenozoic, Ch. 3. In: Whitmore, T.C. (Ed.). Wallace's line and plate tectonics. Clarendon Press, Oxford. 91 pp.

Babcock, J.P. 1914. The spawning beds of the Fraser. Rept. B.C. Comm Fish. 1913: 17-38.

Barash, D. 1979. The whisperings within: evolution and the origin of human nature. Harper and Row, N.Y. 274 pp.

Barnett, S.A. 1963. A study in behavior: principles of ethology and behavioral physiology displayed mainly in the rat. Methuen, London. 288 pp.

Berger, J.R. 1977. Northern frontier. Northern homeland. Rept. of the Mackenzie Valley Pipeline Inquiry. 2 vols. Min. Supply and Services, Ottawa.

Bishop, J.A. and L.M. Cook. 1975. Moths, melanism and clean air. Sci. Am. 232(1): 90-99.

Bjorndal, K. 1981. Biology and conservation of sea turtles. Proc. World Conf. Sea Turtle Cons. Smithsonian Institution Press, Washington, D.C. 583 pp.

Blakemore, R.P. and R.B. Frankel. 1981. Magnetic navigation in bacteria. Sci. Am. (Dec.): 58-65.

Breland, K. and M. Breland. 1966. Animal Behavior. MacMillan, N.Y. 210 pp.

Bull, H.O. 1957. Conditioned responses, Ch. 3, Part 1. In: Brown, H.E. The physiology of fishes, Vol. II: Behavior. Academic Press, N.Y.

Calhoun, J.B. 1962. The ecology and sociology of the Norway rat. U.S. Dept. Health, Education and Welfare, Bethesda, MD. 288 pp.

Carpenter, C.R. 1934. A field study of the behavior and social relations of howling monkeys. John Hopkins Press, Baltimore. Comp. Psych. Monographs, 10(2) Serial No. 48. 168 pp. Reprinted 1967. Kraus Reprint Corporation.

Carr, A. and P.J. Coleman. 1974. Seafloor spreading theory and the odyssey of the green turtle. Nature 249: 128-30.

Chalmers, N. 1979. Social behavior in primates. Arnold, London. 256 pp.

Charles-Dominique, P. and R.D. Martin. 1972. Behaviour and ecology of nocturnal prosimians. Fortschr. Verhaltensforsch. 9: 7-90.

Clark, R.B. 1964. Dynamics in metazoan evolution. Clarendon, Oxford. 313 pp.

Clay, C.H. 1961. Design of fishways and other fish facilities. Dept. Fisheries, Canada, Ottawa. 301 pp.

Clements, F.E. and V.E. Shelford. 1979. Bio-ecology. Wiley, London. 425 pp.

Collins, R.L. and J.L. Fuller. 1968. Audiogenic seizure prone (asp): a gene affecting behavior in linkage group VIII of the mouse. Science 162: 1137-1139.

Coon, C.S. 1965. The living races of man. Knopf, N.Y. 344 pp.

Craig, W. 1918. Appetites and aversions as constituents of instincts. Biol. Bull. 34: 91-107.

Dane, B., C. Walcott and W.H. Drury. 1959. The form and duration of the display actions of the goldeneye (*Bucephala clangula*). Behaviour 14: 265-281.

Darwin, C. 1839. Journal of researches into the geology and natural history of the various countries visited by H.M.S. Beagle. Henry Colburn, London. (1961 Dent: Everyman's Library, London).

Darwin, C. 1859. On the origin of species. John Murray, London. (1964 Facsimile edition, Harvard University Press, Cambridge, MA).

Darwin, C. 1868. The variation of animals and plants under domestication. John Murray, London. (1905 revised edition, F. Darwin, Ed).

Darwin, C. 1872. The expression of the emotions in man and animals. (1965 University of Chicago reprint, 372 pp.).

Darwin, C. 1881. The formation of vegetable mould, through the action of worms. John Murray, London. (1945 "Darwin on humans and the earthworm", Faber and Faber, London).

Davidge, C. 1976. Activity patterns of baboons (*Papio ursinus*) at Cape Point. M.Sc. Thesis, University of Cape Town.

Dawkins, R. 1976. The selfish gene. Oxford University Press, Oxford. 224 pp.

Dawkins, R. and J.R. Krebs. 1978. Animal signals: information on manipulation. Ch. 10, pp. 282-315 in: Krebs, J.R. and N.B. Davies (Eds.). Behavioural ecology: an evolutionary approach. Blackwell, Oxford. 494 pp.

DeFries, I.C. and C.E. McLearn. 1970. Social dominance and Darwinian fitness in the laboratory mouse. Am. Nat. 408-411.

Desmond, A.J. 1975. The hot-blooded dinosaurs. Blond and Briggs, London. 238 pp.

Dilger, W.C. 1960. The comparative ethology of the African parrot genus *Agapornis*. Zeitschrift fur Tierpsychologie 17: 649-685.

Dilger, W.C. 1964. The interaction between genetic and

experimental influences in the development of species-typical behavior. American Zoologist 4: 155-160.

Douglas-Hamilton, I. and O. Douglas-Hamilton. 1975. Among the elephants. Collins and Harvill Press, London.

Eibl-Eibesfeldt, I. 1970. Ethology: the biology of behavior. Holt, Rinehart and Winston, N.Y. 530 pp.

Eisenberg, J.F., G.M. McKay and M.R. Jainudeen. 1971. Reproductive behavior of the Asiatic elephant (*Elephas maximus maximus* L.). Behavior 38: 193-225.

Elliott, I.M. 1979 (Second printing, second edition). Statistical analysis of samples of benthic invertebrates. Freshw. Biol. Assoc. Sci. Publ. 25. 160 pp.

Ellis, D.V. 1966a. Swimming speeds of sockeye and coho salmon on spawning migration. J. Fish. Res. Bd. Can. 23(2): 181-187.

Ellis, D.V. 1966b. A survey of the behavior of salmon on spawning migration through a large river system. Fish. Res. Bd. Can. Mss. Rept. Series (Biological) No. 876. 17 pp.

Ellis, D.V. 1977. The fish: an ethogram for management. Ch. 2, pp. 35-67 in: Ellis, D.V. (Ed.). Pacific salmon: management for people. Western Geographic Series vol. 13. University of Victoria, Victoria, B.C. 320 pp.

Ellis, D.V. 1982. Subordinate sex. Arlington, London. 252 pp.

Elton, C. 1942. Voles, mice and lemmings. Clarendon Press, Oxford. 496 pp.

Fabricius, E. 1950. Heterogenous stimulus summation in the release of spawning activities in fish. Rept. Inst. Freshwater Res. Drottningholm No. 31: 57-99.

Fagen, R. 1981. Animal play behavior. Oxford University Press, Oxford. 684 pp.

Fisher, H.E. 1982. The sex contract. Morrow, N.Y. 253 pp.

Fitzpatrick, S.M. and W.G. Wellington. 1983. Insect territoriality. Can. J. Zool. 61(3): 471-486.

Ford, J. and D. Ford. 1981. The killer whales of B.C. waters. Journal of the Vancouver Aquarium 5(1): 3-31.

Fox, R. 1980. The red lamp of incest. Dutton, N.Y. 271 pp.

Fraenkel, G.S. and D.L. Gunn. 1961. The orientation of animals. Dover, N.Y. 376 p. (Reprint of original edition - 1940. Oxford University Press.)

Frisch, K. von. 1953. The dancing bees. Harcourt, Brace and World, N.Y. 182 p. (First English edition).

Fuller, J.L. and M.W. Fox. 1969. The behavior of dogs. Ch. 14, pp. 438-481 in: Hafez, E.S.E. (Ed.). The behaviour of domestic animals. Bailliere, Tindall and Cassell, London. 647 pp.

Gardner, M. 1981. Science: good, bad and bogus. Avon, N.Y. 408 pp.

Gerking, S.D. 1959. The restricted movement of fish populations. Biological Reviews 34: 221-242.

Gittelsen, B. 1975. Biorhythm: a personal science. First edition. Warner, N.Y. 407 pp.

Golding, W. 1954. Lord of the flies. Faber and Faber, London. 248 pp.

Golding, W. 1955. The inheritors. Faber and Faber, London. 282 pp.

Goodall, J. 1979. Life and death at Gombe. Nat. Geog. Mag. 155(5): 592-620.

Gould, J.L. 1982. Ethology: the mechanisms and evolution of behavior. Norton, N.Y. 544 pp.

Gowaty, P. 1982. Sexual terms in sociobiology: emotionally evocative and, paradoxically, jargon. Animal Behavior 30: 630-631.

Gribbin, J. and J. Cherfas. 1982. The monkey puzzle. Bodley Head, London. 280 pp.

Griffin, D.R. 1958. Listening in the dark. Yale University Press, New Haven. 413 pp.

Groot, C. 1965. On the orientation of young sockeye salmon (*Oncorhynchus nerka*) during their seaward migration out of lakes. E.J. Brill, Leiden. 198 pp.

Hailman, J.P. 1969. How an instinct is learned. Sci. Am. 221(6): 98-106.

Hale, E.B. 1969. Domestication and the evolution of

behavior. Pp. 22-24 in: Hafez, E.S.E. (Ed.). The behaviour of domestic animals. Balliere, Tindall and Cassell, London. 647 pp.

Halstead, B.W. 1965. Poisonous and venomous marine animals of the world. U.S. Govt. Printing Office, Washington, D.C. 3 vols.

Harden-Jones, F.R. 1968. Fish migration. Arnold, London. 325 pp.

Hardy, A. 1960. Was man more aquatic in the past? New Scientist 7: 642-645.

Harlow, H.F. and M.K. Harlow. 1962. Social deprivation in monkeys. Sci. Am. 207: 137-146.

Hay, R.L. and M.D. Leakey. 1982. The fossil footprints of Laetoli. Sci. Am. 246(2): 50-57.

Hediger, H. 1950. Wild animals in captivity. Butterworth, London. 207 pp.

Hitler, A. 1939. Mein Kampf. English translation. Houghton-Mifflin, Boston. 1003 pp.

Hoar, W.S. 1965. The endocrine system as a chemical link between the organism and its environment. Trans. Roy. Soc. Can. Fourth Series. Vol. III, Section 3: 175-200.

Huxley, J.S. 1914. The courtship habits of the great crested grebe (*Podiceps cristatus*) with an addition on the theory of sexual selection. Proc. Zoo. Soc. London 35: 491-562.

Huxley, J.S. and H.B.D. Kettlewell. 1965. Charles Darwin and his world. Viking, N.Y. 144 pp.

Jennings, H.S. 1976. Behavior of the lower organisms. Indiana University Press, Bloomington. 366 pp. (Originally published 1905.)

Johanson, D.C. and M.A. Edey. 1981. Lucy: the beginnings of humankind. Simon and Schuster, N.Y. 409 pp.

Johnson, C.G. 1969. Migration and dispersal of insects by flight. Methuen, London. 263 pp.

Johnson, L.C., W.P. Colquhoun, D.I. Tepus and M.J. Colligan (Eds.). 1981. Biological rhythms, sleep and shift work. Adv. Sleep Research Vol. 7. SP Medical and Scientific Books, N.Y. 618 pp.

Kettlewell, B. 1973. The evolution of melanism. Oxford

University Press, Oxford. 423 pp.

Kettlewell, H.B.D. 1959. Darwin's missing evidence. Sci. Am. 200(3): 48-53.

Kortlandt, A. 1972. New perspectives on ape and human evolution. Stichting voor Psychobiologie, Amsterdam.

Koshtoyants, K.S. (ed.). No date. I.P. Pavlov. Selected works. Foreign Language Publishing House, Moscow. 662 pp.

Kuo, Z-Y. 1967. The dynamics of behavior development. Random House, N.Y. 240 pp.

Kurten, B. 1970. Continental drift and evolution. Ch. 12 in: Wilson, J.T. (Ed.). Continents adrift. Freeman, San Francisco.

Lack, D. 1956. Swifts in a tower. (not seen).

La Lumiere, L.P. 1981. Evolution of human bipedalism: a hypothesis about where it happened. Phil. Trans. Roy. Soc. London B292: 103-107.

Lawn, I.D., G.O. Mackie and G.A. Silver. 1981. A conduction system in a sponge. Science 211: 1169-1171.

Lockyer, C.H. and S.C. Brown. 1981. The migration of whales. Pp. 105-137 in: Aidley, D.J. (Ed.). Animal migration. Soc. Exp. Biol. Sem. Series 13. Cambridge University Press, Cambridge. 264 pp.

Longhurst, A.R. 1976. Vertical migration. Ch. 6, pp. 116-140 in: Cushing, D.H. and J.J. Walsh (Eds.). The ecology of the seas. Saunders, Philadelphia. 467 pp.

Lorenz, K. 1937. The companion in the bird's world. Auk. 54: 245-273.

Lorenz, K. 1950. The comparative method in studying innate behavior patterns. Symp. Soc. Exp. Biol. 4: 221-268.

Lorenz, K. 1951. Comparative studies on the behavior of Anatinae. Avicultural Magazine 57:157-182; 58 (1952): 8-17, 61-72, 86-94, 172-184; and 59 (1953): 24-34, 80-91.

Lorenz, K. 1952. King Solomon's ring: new light on animal ways. Crowell, N.Y. 208 pp.

Lorenz, K. 1955. Man meets dog. Houghton Mifflin, Boston. 211 pp.

Lorenz, K. 1958. The evolution of behavior. Sci. Am. 199(6): 67-78.

Lorenz, K. 1966. On aggression. Harcourt, Brace and World, N.Y. 306 pp.

MacArthur, R.H. and E.O. Wilson. 1967. The theory of island biogeography. Princeton University Press, Princeton, N.J. 203 pp.

McConnell, J.V. 1974. Understanding human behavior: an introduction to psychology. Holt, Rinehart, N.Y. 831 pp.

McGrath, T.A., M.D. Shalter, W.M. Schleidt and P. Sarvella. 1972. Analysis of distress calls of chicken x pheasant hybrids. Nature 237: 21-27.

McKervill, H.W. 1967. The salmon people. Gray's Publishing Co., Sidney, B.C. 198 pp.

Mackie, G.O. 1970. Neuroid conduction and the evolution of conducting tissue. Quarterly Review of Biology 45(4): 319-332.

Mackie, G.O., I.D. Lawn and M. Pavans de Ceccatty. 1983. Studies on hexactinellid sponges. II. Excitability, conduction and coordination of responses in *Rhabdocalyptus dawsoni* (Lambe. 1873). Phil. Trans. Royal Soc. London. B. Biol. Sciences: 301(1107): 401-418.

Marais, E. 1969. The soul of the ape. Atheneum, N.Y. 226 pp.

Marais, E.N. 1971. The soul of the white ant. Jonathan Cape and Anthony Blond, London. 139 p. (Originally published in Afrikaans in 1937, and based on field studies made largely from 1910-1915).

Markowitz, H. and V.J. Stevens (Eds.). 1978. Behavior of captive wild animals. Nelson-Hall, Chicago. 314 pp.

Martin, R.D. and R.M. May. 1982. Outward signs of breeding. Pp. 234-239 in: Maynard Smith, J. Evolution now. Freeman, San Francisco. 239 pp.

Maynard Smith, J. 1975. The theory of evolution. Third edition. Penguin Books, London. 344 pp.

Maynard Smith, J. 1978. The evolution of behavior. Sci. Am. 239(3): 176-193.

Meddis, R. 1975. On the function of sleep. Animal

Behavior 23: 676-691.

Mellen, S.L.W. 1981. The evolution of love. Freeman, San Francisco. 312 pp.

Mitchell, B. 1977. Hindsight reviews: the British Columbian licence programme. Ch. 7, pp. 148-186 in: Ellis, D.V. (Ed.). Pacific salmon: management for people. Western Geographical Series No. 13. University of Victoria, Victoria, B.C. 320 pp.

Morgan, E. 1972. The descent of woman. Souvenir Press, London. 288 pp.

Morgan, E. 1982. The aquatic ape. Souvenir Press, London. 168 pp.

Morgan, E. 1984. The aquatic hypothesis. New Scientist 1405: 11-13.

Morris, D. 1966. The biology of art. Methuen, London. 176 pp.

Morris, D. 1967. The naked ape. Jonathan Cape, London. 252 pp.

Morris, D. 1969. The human zoo. Clarke-Irwin, Toronto. 256 pp.

Morris, D. 1971. Intimate behaviour. Cape, London. 253 pp.

Morris, D. 1977. Manwatching. Cape, London. 320 pp.

Morris, D. 1979. Animal Days. Book Club Associates, London. 275 pp.

Morris, D., P. Collett, P. Marsh and M. O'Shaughnessy. 1981. Gestures: their origins and distribution. Triad Granada, London. 296 pp.

Mott, P., F. Mann, Q. McLaughlin and D. Warwick. 1965. Shift work: the social, psychological and physical consequences. University of Michigan Press, Ann Arbor, Michigan. 351 pp.

North, W.J. and M.B. Schaefer. 1964. An investigation of the effects of discharged wastes on kelp. Resources Agency of California State Water Quality Control Board. Publication No. 26. Sacramento, California. 124 pp.

Norton, I.O. and J.G. Sclater. 1979. A model for the evolution of the Indian Ocean and the breakup of Gondwanaland. Journ. Geophys. Res. 84(12): 6803-6830.

Packard, J.M. and C.A. Ribic. 1982. Classification of the behaviour of sea otters. Can. J. Zool. 60: 1302-1323.

Palmer, J.D. 1975. Biological clocks of the tidal zone. Sci. Am. 232(2): 70-79.

Parker, G.H. 1919. The elementary nervous system. Lippincott, Philadelphia.

Passingham, R. 1982. The human primate. Freeman, San Francisco. 390 pp.

Pavlov, I.P. 1928. Lectures on conditional reflexes. (English translation - Gantt, W.H. and G. Volbarth (eds.). International Publishers, N.Y. 414 pp.).

Pilbeam, D. 1984. The descent of hominoids and hominids. Scientific American 250(3): 84-96.

Pycroft, W.P. 1913. The courtship of animals. Hutchinson, London.

Rensberger, B. 1977. The cult of the wild. Anchor, N.Y. 279 pp.

Robertson, D.R. 1972. Sex reversal in a coral-reef fish. Science 177: 1007-1009.

Roesijadi, C. 1981. The significance of low molecular weight, metallothionein-like proteins in marine invertebrates: current states. Marine Environmental Research 4: 167-179.

Rohwer, S. and F.C. Rohwer. 1978. Status signalling in Harris sparrows: experimental deceptions achieved. An. Behav. 26: 1012-1022.

Rothenbuler, W.C. 1964. Behavior genetics of nest cleaning in honey bees. American Zoologist 4: 111-123.

Ruse, M. 1979. Sociobiology: sense or nonsense? Reidel, Holland. 231 pp.

Schjelderup-Ebbe, T. 1935. Social behavior of birds. Pp. 947-972 in: Murchison, C.A. (Ed.). A handbook of social psychology. Clark University Press, Worcester, Mass. 1195 pp.

Schmidt-Nielsen, K. 1979. Animal psychology: adaptation and environment. Cambridge University Press, Cambridge. 560 pp.

Schrage, C. No date. The training and care of a killer whale. Sealand of the Pacific, Victoria, B.C. 24 pp.

Scrivener, C. 1970. Agonistic behavior of the American lobster *Homarus americanus* (Milne-Edwards). M.Sc. Thesis, University of Victoria. (1971. Fish. Res. Bd. Can. Tech. Rept. #235, 128 p.)

Simons, E.L. 1964. The early relatives of man. Ch. 2 in: Isaac, G. and R.E.F. Leakey. Human Ancesters. Freeman, San Francisco. 130 pp.

Skinner, B.F. 1938. The behavior of organisms. Appleton-Century Crofts, N.Y. 457 pp.

Strum, S.C. 1975. Life with the pumphouse gang. Nat. Geog. Mag. 147(5): 671-691.

Stuart, T.A. 1964. The leaping behavior of salmon and trout at falls and obstructions. Freshwater and Salmon Fish. Res. No. 28. Dept. Ag. and Fish., Scotland. 46 pp.

Terrace, H.S. 1979. Nim: a chimpanzee who learned sign language. Washington Square Press, N.Y. 437 pp.

Thompson, W.F. 1945. Effect of the obstruction at Hell's Gate on the sockeye salmon of the Fraser River. Int. Pac. Salm. Fish. Comm. Bull. 1. 175 pp.

Tiger, L. 1971. Men in groups. Random House, N.Y. 254 pp.

Tinbergen, N. 1942. An objective study of the innate behavior of animals. Biblioth. biotheor. 1: 39-98.

Tinbergen, N. 1951. The study of instinct. Oxford University Press, Oxford.

Tinbergen, N. 1960. The evolution of behaviour in gulls. Sci. Am. 203(6): 118-130.

Tinbergen, N. and J.J.A. van Iersal. 1947. "Displacement reactions" in the three-spined stickleback. Behaviour 1: 56-63.

Todd, I.S. 1969. The selective action of gillnets on sockeye (*Oncorhynchus nerka*) and pink salmon (*O. gorbuscha*) stocks of the Skeena River system, British Columbia. M.Sc. thesis, University of British Columbia. 141 pp.

Van Lawick, H. and J. van Lawick-Goodall. 1970. Innocent killers. Collins, London. 222 pp.

Van Lawick-Goodall, J. 1971. In the shadow of man. Collins, London. 287 pp.

Vogel, S. 1978. Organisms that capture currents. Sci. Am. 239(2): 128-139.

Webb, J.E. 1976. A review of swimming in Amphioxus. Pp. 447-454 in: Davies, P. (Ed.) Perspectives in experimental biology, vol. 1: Zoology. Proc. 50th Anniversary Meeting Society Exp. Biol., Pergamon.

Wells, M.J. 1978. Octopus. Chapman and Hall, London. 417 pp.

Wiley, E.O. 1981. Phylogenetics. Wiley-Interscience, N.Y. 439 pp.

Wilson, E.O. 1970. Chemical communication within animal species. Ch. 7, pp. 133-135 in: Sandheimer, E. and J.B. Simeone (Eds.). Chemical ecology. Academic Press, N.Y. 336 pp.

Wilson, E.O. 1975. Sociobiology. Harvard University Press, Cambridge, Mass. 697 pp.

Wilson, E.O. 1978. On human nature. Harvard University Press, Cambridge, Mass. 260 pp.

Woese, C.R. 1981. Archaebacteria. Sci. Am. 244: 98-122.

Wu, R. and Lin, S. 1983. Peking man. Sci. Am. 248(6): 86-94.

Wynne-Edwards, V.C. 1962. Animal dispersion in relation to social behavior. Oliver and Boyd, Edinburgh. 653 pp.

Zuckerman, S. 1932. The social life of monkeys and apes. Inst. Lib. Psych. Phil and Sci. Method. Kegan Paul, London. 357 pp.

APPENDIX 1. THE ETHOGRAM AS A CHECKLIST.

The following is provided as a checklist of behaviors to be looked for in a species.

I. Obtaining materials and energy.

A. Feeding

1. Herbivore - browsing, grazing?
2. Carnivore - ambush (drift or hide), run down?
3. Omnivore - random, selective?
4. Particle feeder - suspension, deposit, current capturing?
5. Scavenger - opportunistic or obligate?

B. Drinking - occurs or not? if not, diffusion or desalination?

C. Breathing - air or water? social signalling? rate constant or variable?

D. Defecation - rate? social signalling? location constant or random?

E. Urination - combined with defecation or separate? social signalling?

II. Avoiding becoming materials and energy for some other animal

A. Escape - alert behavior? dart away or sudden withdrawal? group explode or cohere? alarm calls or alarm flashes? aversive or cautious behavior by predator? self-defense?

B. Hide - habitat or biological screen? freeze? cryptic coloration?

C. Return - cautious? sensing?

D. Wounded - extended hiding? antiseptic behavior?

III. Maintaining the right habitat

A. Habitat - physical, chemical or biological? structures or featureless?

B. Intra-habitat restrictions - territory, home range or both? individual or group occupation? home range sharing simultaneous or sequential? personal space? alert, flight and defended distances?

C. Limiting factors - avoidance behavior?

D. Optimal conditions - location behavior?

E. Habitat sequencing - periodic movements? local or distant movements? short or long time-scales? life cycle habitat changes?

IV. Inactivity

A. Reduced metabolism - overt signs? slow breathing? risk avoiding behavior?

B. Fatigue - duration of a few seconds or minutes? preceded by violent activity? fast breathing?

C. Sleep - duration of a few minutes to hours? periodic? regular? daytime? nocturnal? alternating high activity? at dawn or dusk? or other time? insulation behavior? social withdrawal? postural behavior? REM sleep?

D. Waiting - duration of seconds to hours? alert? sensors deployed? immobile or slow ranging?

E. Hibernation - duration of weeks to years? season? risk reducing behavior? next generation benefit?

F. Invertebrates - any other associated behaviors? e.g. molting?

V. Removing disturbing stimulii (comfort movements)

A. External - scratch? rub? social signalling? symbiotic behavior?

B. Internal - yawn, cough, or sneeze? associated chest or operculum movements? associated head movements? other channels flushed?

VI. Perpetuating oneself

A. Asexual? sexual?

B. Courtship - present or not? external fertilization? pheromones or mutual environmental sensitivity? large animal? navigation problems? fierce animal? aggression suppression? sessile (sedentary)? external or internal fertilization? mixed sex clusters or single sex? habitat selection? mate selection?

C. Mating - quick? extended? repeated?

D. Parental behavior - young in right habitat? habitat changes? parental protection of young? by structures? or by behavior? parental guidance to young? opportunities to learn? teaching by example? grandparenting?

VII. Social organization

A. Social behavior - group formation? permanent? coherent? outsiders? organizational behavior? non-organized aggregation?

B. Limiting resources? food? mates? space? grooming rights? other?

C. Organizational tactics

1. Ranking - dominant? leader or initiator? linear or looped? male and female relative ranks? individuals or alliances? group rankings? outright aggression or covert threats?
2. Territory - defense by aggression or threat? male or female defense or both? group territory? or home range? one or more territories? different resources? personal space?
3. Castes - types? feeding? breeders? warriors? drones? nurses? others? what activities are common to castes? comfort? carrying?

D. Communication

1. Visual signals or badges? or both?
2. Vocalizations? calls? songs? language?
3. Smells? pheromones? marking? body sniffing? defense?
4. Other sensory modalities? electrical signalling?

E. Competition/cooperation/altruism/spite

1. Can any behaviors be labelled by these terms?
2. Do they draw on categories already in the checklist? e.g. attack, feeding, alert?
3. Is the group comprised of kin? how closely related? can you recognize individuals?

F. Intraspecies differential behavior - parent-offspring conflicts? older-younger sibling conflicts? male-female conflicts?

G. Interspecific cooperative behavior - mutualism, commensalism or parasitism? searching or avoidance behavior? life cycle habitat transfers?

APPENDIX 2. A SUMMARY ETHOGRAM FOR A SELECTED SPECIES GROUP (THE PACIFIC SALMONS) WITH IMPLICATIONS FOR MANAGEMENT BIOLOGISTS (modified from Ellis 1977).

It is important that the ethogram concept be applied in practice, either by the novice behaviorist making personal observations on a particular species or by reviewing literature on closely related species, or both. An example is now given for Pacific salmon in western North America (Table A2.1). The generalized life cycle is summarized in Figure A2.1. This example ethogram is expressed in the context of management biologists' needs to relate their specific interests, say in gear design or escapement control, to the total behavior of the species.

I. Obtaining materials and energy

Pacific salmon illustrate a common materials gathering strategy, that of changing their food source with growth. Initially the embryo, within the egg, as it grows through the alevin stage lying in the gravel, is dependent on the yolk laid down by the mother. Soon after leaving gravel for open water, the fry start independent food gathering. (This may be delayed for a day or so for pink and chum fry. They make a nightly non-feeding journey down to the estuary, burrowing back into gravel during daylight.) The food gathering at this stage consists of opportunistic feeding on virtually any small organic matter, live or dead, which drifts, swims, or even flies past. Such tactics imply that the small 2 cm long fish have already developed acute three-dimensional vision sufficient to take an accurate fix on moving food, the coordination to intercept the drift and snap it up in motion, and the sensitivity to reject the material if not edible. There is even more to their capability since the continual supply of drift food from upstream can be best tapped by individual fish maintaining a fixed station. By returning to such a station after each feeding foray they are prevented from unwittingly drifting downstream. As a result aggressive territorial competition between individuals for prime available space, i.e. good feeding stations, sets in.

Table A2.1. Pacific salmon species in North America.

Common name	**Scientific name**	**Special features of social behavior and habitats**
Pink (humpback) salmon	*Oncorhynchus gorbuscha*	2 year life cycle Gravel to estuary migration Estuary - first spring Return to spawn - 2nd summer
Chum (dog) salmon	*Oncorhynchus keta*	3-5 year life cycle Gravel to estuary migration Several years at sea
Coho (silver) salmon	*Oncorhynchus kisutch*	3-4 year life cycle River territories as juveniles River to estuary migration at start of 2nd year as smolts Can spawn in stream headwaters - migration can be in pairs Jack males occur
Red (sockeye) salmon	*Oncorhynchus nerka*	4-5 year life cycle Lake residence as juveniles Lake to estuary migration as 2+ year smolts
Chinook (spring, king) salmon	*Oncorhynchus tshawytscha*	3-8 year life cycle River territories as juveniles River to estuary migration from end of 1st summer May grow very large Jack males occur

The fry of species which quickly enter large bodies of water, the pinks, chums and sockeyes, feed there in a similar way at first but are not exposed to a unidirectionally flowing habitat. Instead the water medium may be still, or virtually so, or possibly influenced by more or less regularly changing tidal currents. The effects of this on feeding tactics are that the fish must actively seek their food rather than have it come to them. Thus fry in lakes and estuaries can commonly be seen wandering in schools through the shallows, a behavior which balances food gathering in a productive environment with protection from underwater predators (but not, of course, from kingfishers and other birds overhead).

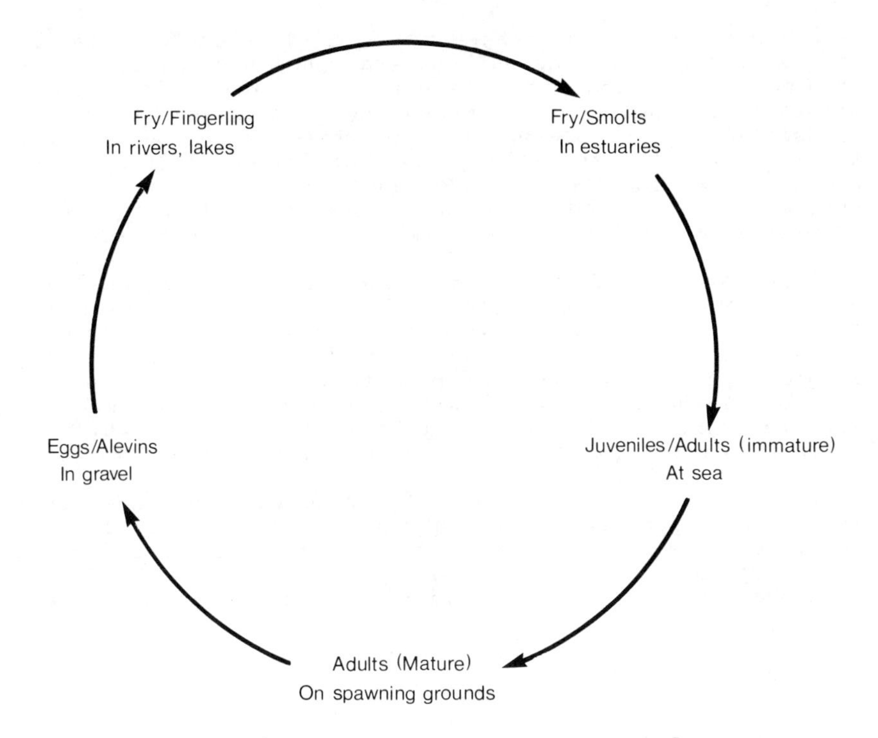

Figure A2.1. Generalized life cycle of the Pacific salmons.

Once the juveniles move offshore to lake depths or coastal seas, and then to the ocean, our knowledge of their feeding behavior is more limited since their actions can rarely be observed directly. It is known that they become true predators in that stomach contents show their prey at sea to be smaller fish (for cohos and chinooks) and larger zooplankton or invertebrate nekton, for example small squids (for sockeye, pink and chum). We can speculate that they use the tactics of chasing type predators utilizing organisms smaller than themselves for food. They remain in the top few hundred feet of water (where the food is concentrated), and they appear to travel individually or at most in loose aggregates which may not justify the term schools.

The salmon have evolved a time related feeding strategy fairly common to those animals which have a far ranging life cycle. Prior to the time to start the long return to the breeding grounds, the fish lay down an energy store within their bodies, the amount of which is adapted to their specific needs for the return journey. This last adaptation is a significant management concern since unusual blocks to the upstream migration, whether artificially induced by fishways or naturally induced by low river flows or landslides, can so use up the programmed energy supply that the fish cannot survive long enough to reach their spawning grounds and complete the prolonged activities of mating. Furthermore as feeding stops, lure type fishing gears simulating natural prey eventually become ineffective.

There are a number of environmental circumstances which also favor intermittent or periodic feeding. Food for salmon at sea is often widely dispersed, either through being in the form of small fish and squids ranging actively in groups, or in clusters of plankton. The salmon may need to provide themselves with sufficient energy at one meal to last for some time, and so may feed to satiation, which by definition precludes further feeding activity for a while.

There is a great deal of anecdotal evidence from fishermen supporting the hypothesis of varying feeding activity in salmon. There are times and places where salmon simply appear not to be hungry, and it is the skill of the fisherman which provokes a strike at a lure at these times.

II. Avoiding predation

Pacific salmon avoid predation by a number of strategies. Initially as small fish in streams and estuaries, where they are subject to predation by many other species, they have two escape systems. The first is the emergency response - an explosive, virtually undirected dart-off which is simply too sudden and too fast for the hunting predator to follow. But this must quickly give way, through risk of fatique, to the second response, that of hiding. Hiding places vary with the size of fish and nature of the predator, from a tangle of underwater vegetation, to holes within a gravel bed, river cut banks and even the depths of river channels and pools beyond the reach of shallow feeding gulls and bears.

These two responses imply a great deal about the salmon's sensitivity even at the fry stage. They can recognize a predator, and there is advantage to them in responding positively to stimuli indicating depth or other hiding places. Predators generally can be perceived visually by dark shadow, often approaching fast or at least moving in an alert fashion. This stimulus paradigm appears to have generated one of the problems affecting hatchery reared fish which have often been artificially fed in such a way that large shadows overhead mean food. After release to the wild, such a stimulus situation has the totally diferent meaning of a predator. As a result, hatchery fish have low survival rates in the wild. They have learned the wrong association.

Learning may also be implicated in the hiding response of salmon. Salmon introduced into a new environment, whether it be an aquarium, river pool or fishway, will often engage initially in more or less random wanderings, giving the impression of exploration of, and even curiosity about, their surroundings. Fishway design might well benefit from considering this point and its implications for speeding migration through a repetitive structure for which the basic shape can be learned in the first few pools.

III. Maintaining and changing habitat

It is arguable where the first habitat maintenance patterns arise since the initial wrigglings of alevins in the gravel may be insufficient to take them significantly far from the point at which the eggs were laid. However, the first habitat change is clear. Late in the winter the alevins' wriggling is sufficiently directed that they move upwards, not randomly in the gravel, and eventually they emerge in enormous numbers over a period of a few days to start their free swimming existence.

The next stage for cohos and chinooks clearly involves habitat maintenance. There is considerable survival value to them in maintaining a feeding station at a point where much food will drift by. Such underwater stations are usually constant in position relative to visually obvious underwater structures such as boulders and vegetation, and even to derived eddy patterns generating points of low current velocity in which a swimming fish can coast with an energy saving, while being in close proximity to a food-providing current. The best feeding stations are commonly occupied by the largest or the most aggressive individuals, aggressive in the sense that they will attack and nip an intruder of their own or similar species. Unlike many other animals though, the young salmon occupy their territory for only short periods of time, possibly as little as a few days. The fish change station as they grow and develop the strength to maintain a position in faster water where there is more food, but also strength to avoid fiercer predators from which they must escape. They are also subject to displacement as river flows change and may sweep them away, or drop to levels such that previously good stations can no longer support active and voracious feeders.

This habitat maintenance by territoriality is lost when the cohos and chinooks make their next change to the open water environment of lakes, an environment where they have been preceded by sockeyes. This is similar, in some ways, to the estuarine environment which received the pink and chum fry. In open still water there is no competitive advantage to a good feeding station, and the food, though abundant, must be sought out with the predator taking the initiative. However, the risk of predation to small fish is considerable in open water, and the social behavior of schooling, with its advantage of confusing predators, appears.

There appear to be some habitat maintaining behaviors in operation in open water, at least in lakes, since juvenile sockeye are reported as keeping within particular lake basins, and schools of lake juveniles can be seen escaping from the river generated currents which could drift them out of their lakes in advance of the right time to make the next habitat change.

This next habitat shift is the one that takes the freshwater juveniles down to the sea in the form of smolts. On their way, they may pass along the length of one or more lakes, and they do so in a very directed and speedy manner. They have a directional sense which serves to take their schools along the lake out of contact with the shoreline, even if it is irregularly shaped. What is more, the directional sense is time compensated so that the shifts in directional trends while moving through the open water of the lakes coincide with shifts in the axis of the lake. In other words salmon appear to have a biological clock within their bodies.

These apparently inherited directional tendencies and their time compensated changes are significant in management for their importance in determining stocks for transplantation purposes, whether to open up new spawning areas or to regenerate the old and devastated. If they are real and innate phenomena, then transplanted stocks must be chosen for their directional sense, which should correspond to the long axes of any lakes that they will need to pass on migration.

Once down to the sea in the near shore waters schooling, as opposed to territoriality, remains the predominant social behavior. However, information from the high seas fisheries suggests that growing salmon there are not strongly schooled as they feed and wander.

In the high seas, a new habitat maintenance pattern operates. The salmon maintain themselves within certain geographical limits, remaining essentially within the north Pacific subarctic oceanographic domain (Figure 8.4). For some reason, few overlap with tuna to the south in the warmer and more saline subtropical domain and few stocks inhabit the colder and less saline waters of the arctic ocean. Apparently they have some sensitivity to the environment on a geographic scale, or else drift forces serve to keep them within their known bounds. Drift probably has some influence since there are known major ocean gyrals roughly corresponding with the wanderings of particular stocks. On the other hand, the known patterns of movement suggest some orientation by the fish to directional cues from the environment. The mechanism seems to be at an indirect level in the sea, affecting the direction of food gathering wandering only statistically, rather than by the strongly oriented movements which the salmon are capable of when in schools in rivers and lakes. These extensive wanderings create problems for international cooperation in management as stocks of one country wander into areas fished by other nations.

The return home to the spawning ground appears to be started by the salmon's annually cycling clock initiating more directed swimming towards the coast just as the salmon arrive in the spring of the year at the general offshore point of their eventual landfall. Less is known concerning this phase than any other, although it is postulated from fishing practices that the fish may on occasion school up intensively en route to the coast, or at other times not until after they arrive there. Profitable seining is obviously dependent on the size of such schools, where they can be found, and the routes they will follow. In 1958 and 1983, there was considerable consternation among management agencies and fishermen of Canada and the U.S.A. when unusually warm oceanographic conditions coincided with Fraser River sockeye stocks making their landfalls from the ocean far to the north of usual locations and traversing the northern Queen Charlotte Strait between Vancouver Island and the mainland to the Fraser River, rather than going via the international waters of the southern Strait of Juan de Fuca. For the individual seiner there is a premium on foresight in predicting where the landfall of the major runs will be and when, and the best seiners understand the ecosystem of ocean environment and salmon migratory/feeding behavior, either intuitively or inductively from their personal records of many years.

The 1958 Fraser sockeye migrations are a good illustration of the complexity of the biological problems underlying habitat changes in the salmon. Whatever sensitivities, or mixture and sequence of sensitivities, are involved, there is some capability of adjustment to different environmental requirements from generation to generation. Postulated mechanisms for oceanic guidance range, for example, from cues in the sky such as responses to the plane of polarization of light to minute electropotentials between sheering currents. The complexity of the life cycle, however, probably requires considerable redundancy in sensitivities at any one time so that many mechanisms can work as alternatives or in sequence as needed.

From the estuary the adult salmon move upriver by steady swimming in schools, interrupted by the needs to escape from the occasional predator and to pass turbulent, fast flowing and even falling water. The different habitat patterns are modelled in Figure A2.2. The strategy adopted appears to be seeking as great a depth as possible, subject to finding currents of sufficiently low velocity so that energy conserving steady swimming, as opposed to fatigue inducing dart swimming, is effective. Standing waves and eddies may also be utilized for energy conservation when leaping (Figure A2.3) or resting. In other words, the upriver migrants have adopted the strategy of the marathon runner rather than the sprinter. Depth seeking on migration is, of course, a protection against predation which in rivers comes largely from surface feeding mammals: bears and humans. However, the greatest river depth may often contain currents so fast that they exceed the ability of salmon to make progress against them without running the risk of oxygen debt. Under these conditions the salmon must balance protection and migration, and so in the Fraser and a few other rivers thousands and even millions of salmon will follow each other upstream just off the banks, where eddies break up the generally strong currents of the main flow.

Homing is pronounced, with accuracy ranging from the least (may be as low as 50%) in pink salmon to almost all in other species.

After the river ascent, the schooled, sexually mature adults move onto the spawning grounds. There they revert to the territorial way of life as females seek and find suitably aerated gravel of the right mechanical type, and the males compete by aggression and threat displays to establish a dominant role in the hours long ritual movements leading to mating. On the spawning grounds the salmon have to abandon their previous behavior as predatory hunters, only occasionally banding together into schools, to adopt a more direct one-to-one mutually sequencing social behavior which will ensure fertilization of eggs and their placement in the right habitat for the next generation of the stock.

IV. Inactivity

Salmon have periods of inactivity. They can often be seen, by an above surface observer or a scuba diver, remaining quietly in position in pools or eddies. It is easy to separate such resting body action from the alert station holding of a feeding juvenile or spawning fish. Nevertheless, we must be careful in calling such inactivity

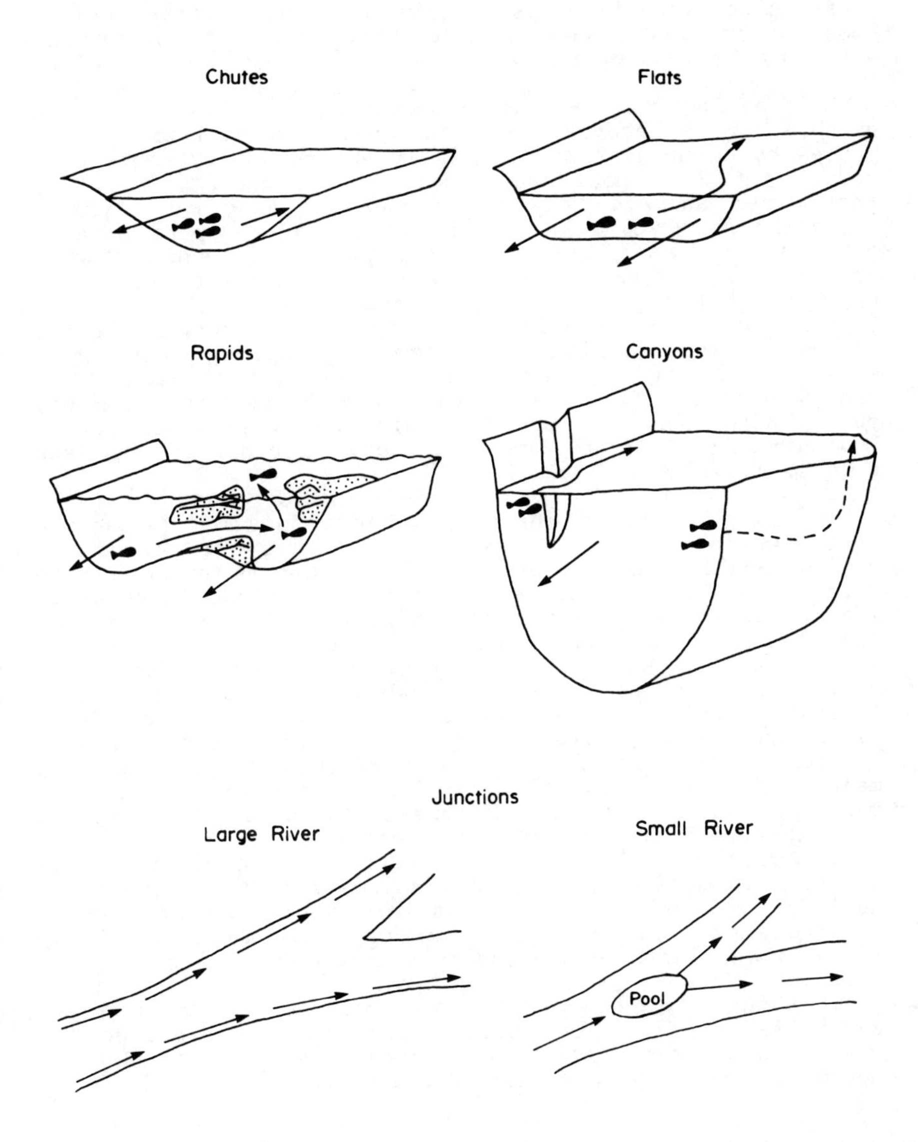

Figure A2.2. Salmon swimming patterns in river habitats.

"rest" since inactivity is relative and the inactive periods come at various stages in the life cycle and occur for various durations of time.

Maintaining stations in an eddy of a falls or rapids appears to be recovery from fatigue. In addition, short pauses by an upriver migrant as it enters a new, restricted river or fishway pool are similar to the cautious scrutiny that other animals take on approach to new and alarming situations. However, migrating salmon often spend several hours a day in a particular river pool, and each day at a certain time, often in the early afternoon, a pool will fill up with a batch of migrants, as though the stock is on a 24 hour migration-rest cycle. This type of inactivity is the closest to what we might recognize as sleep, even though it may occur during daylight hours rather than at night. It should be noted that many birds commonly rest in this way during afternoon as well as at night, and are most active from dawn through the morning, and secondarily in late afternoon to dusk.

There is a longer inactivity period assumed by salmon stocks which hold up in lakes or estuaries for periods of days or weeks before entering spawning streams or their home river respectively. These are actually periods of relative inactivity rather than rest, since the fish may spend considerable time daily cruising the lake or estuary, leaping at the surface as they go and often following an obvious 24 hour cycle. Such periods of inactivity cannot be fatigue, caution or sleep as we know them. These breaks in migration are more likely to be programmed waiting periods, required either before a passage which physically can be made only at a certain part of the season, or prior to making a physiological adjustment, such as the osmotic control capability required for entry to rivers from the sea. Such long waiting periods are features of considerable interest to management and to fishermen, since during these periods the fish may be accessible when conservation is required, or conversely the fish may wait in inaccessible places when fishing is needed to reduce escapements.

The capacity for waiting and resting is specific to particular stocks. It influences management practice and may need identification and consideration in whatever management decisions are made for any stock.

V. Removing disturbing stimulii

Salmon, like other animals, engage in actions which to our human interpretation look as though they are functional in caring for the body. For example they cough, yawn, and rub themselves against a solid object. They even shake their heads spasmodically in a manner similar to human sneezing. Since coughs and sneezes often occur when drift matter is obvious in the water, it is reasonable to conclude that they serve the same function in fish as in people. Yawning in humans flushes out lungs with air, and in salmon can be seen to coincide with a violent jet of water squirted out under the gill covers.

Salmon also engage in a number of apparently useless actions, or actions the function of which is arguable.

Salmon leaping in open water is perhaps the most obvious of the controversial actions. No one argues that salmon leaping at falling water is not functional and the energy

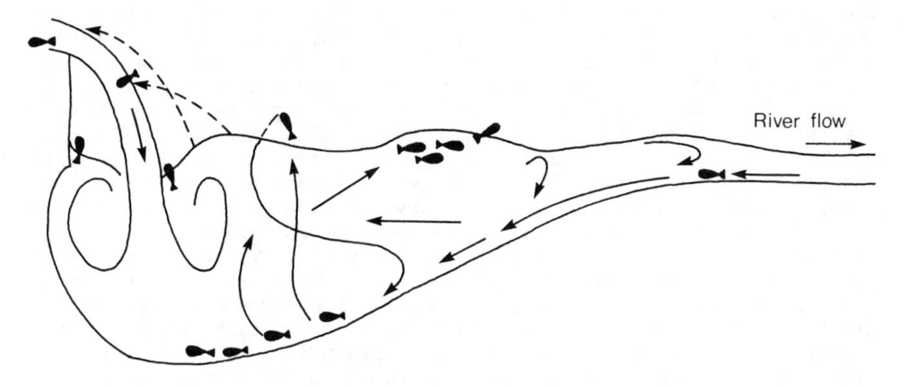

Figure A2.3. Model of salmon leaping behavior and conditions under which leaping occurs.

benefits have been well described (Figure A2.3). However salmon also leap at sea, in estuaries, in river pools, and in lakes at the point of entry from rivers and the point of exit to spawning streams. Commonly it is believed that this shakes the roe loose, or removes parasites. Alternatively, salmon leaping in open water could be an irrelevant action triggered due to the simultaneous conflict between stimulii releasing migration and inhibiting it, i.e. a displacement reaction to incompatible stimulii.

When irrelevant actions are numerous along the route of a migrating stock, there may be a biological problem which could be resolved by appropriate recognition and management action. As so many of the irrelevant actions relate to care of the body, and the alternative of displacement theory indicates stressful situations, either postulate means that it is good management to investigate the causes of any massed or intensive inappropriate actions.

VI. Perpetuating oneself

The Pacific salmons all pass through an extended courtship behavior showing minor modifications on the general model in Table A2.2. Essentially, the female locates a section of gravel providing movable stones and pebbles and having good aeration. She establishes a territory which she will defend against other females. Males congregate around her and fight by biting, which is directed largely at caudal peduncles. They may threaten by swimming up to five meters or so, parallel to one another with raised dorsal fins. Eventually one dominates and subordinates may remain to the rear and sides on convenient energy-assist standing waves. The dominant male may maintain attacks on them as he proceeds through his courtship. He takes on a dominance skin pattern comprising variations on a theme of gaudy, vertically barred, coloring, while females and subordinate males tend towards a dark lateral stripe. The female digs by turning onto her side, powerfully beating the tail upward and rearward on the gravel and throwing the stones downstream. Once she is well advanced and a pit appears in the gravel, the male starts to approach from behind and vibrates his head at approximately the level of her vent. He also maintains station close behind and above, and crosses over from one side to the other, often allowing ventral and paired fins to trail over her back. There are many visual, vibratory and touch stimulii possible within these complexes of displays, especially as the eyes of salmon are located where they have a three-dimensional vision to the rear and upward, just where the male may attend the female.

As the redd is dug out, the female breaks into arcing behavior in which she drops her ventral fin into the redd. The arc taken by her back presumably serves the function of feeding information back to her brain about the depth of the redd and its suitability to receive eggs. Meanwhile the male vibrates progressively closer to her head than to her vent.

Interspersed with these actions, the female may often rise to the surface apparently to take in and expel air, and thereby adjust her body density to be heavy enough to allow her to dig easily. Often bubbles of air can be seen escaping from her mouth and gills as she descends to the bottom.

Table A2.2. The courtship sequence of Pacific salmon.

Phase	Female	Dominant Male
1	Searches gravel Establishes territory over gravel	Searches gravel Establishes territory around female
2	Digs nest (redd)	Establishes dominance over other males Takes on dominance color patterns
3	Continues digging	Starts courting female Crossovers and vibrates (to vent of female)
4	Tests nest depth - arcing	Vibrates (to head of female)
5	Gaping	Gaping
6	Spawning	Spawning
7	Covering	Resting
1 (Repeat)	Digs new nest ahead of old	Reestablishes dominance or is displaced

Note: Behavior of subordinate males and jacks is not well documented but may give rise to their spawning simultaneously with the dominant.

Eventually on a deep arc the female will gape, vibrate her whole body and release eggs. If the male is properly timed he will swim alongside, tilt and S-shape himself slightly, gape, vibrate and release milt. Both clouds of gametes settle to the gravel redd due to the close placement of fish vents into the water flow boundary layer. One or two subordinate males or jack salmon may swim in and also drop sperm without generating attacks.

After a brief pause, the female breaks into covering behavior somewhat resembling digging. She gently beats her tail against the front edge of the redd so that small stones are rolled back over the fertilized eggs. She progresses into another round of digging so that heavier stones are dropped back and cover the earlier batch of eggs. The male restarts dominance actions, although he may be replaced by another.

The sequence may be repeated several times, until the female's eggs have all been dropped. She then may guard her trench of eggs for a while, until increasing tissue hystolysis, which has provided the energy for spawning after stored reserves have been utilized, progresses to a point where she is unable to resist the current any longer. The final abandonment of territory is also influenced by infection from various fungi and attacks from gulls, bears, and human predators.

During the courting and breeding sequence the salmon may be disturbed by predators and will dart away, sometimes into hiding, and may remain there until darkness before returning to restart the sequence. Otherwise spawning may continue day and night.

The sexes are only clearly separable morphologically during the spawning period, which may present difficulties in escapement calculations for early running stocks, prior to their taking on secondary sexual characters. The management considerations arising from spawning behavior are the area of gravel needed, the physical nature of gravel and its aeration, the extent to which excessive escapement will dig over eggs already laid, whether early or late run escapements have worse spawning success than do season run fish, and the elimination of disturbance from predators, false stimulii, and subordinates during spawning.

VII. Social organization

Three forms of social organization are adopted at various life stages: territoriality, ranking and schooling. Territoriality occurs during juvenile stream residence. The young salmon spread themselves through space, balancing needs for energy assists from standing waves, and feeding on drift material. There is fierce competition, and a resident will defend his territory with active attacks on intruders and neighbors. Threat displays consist of fin and operculum raising. The territorial species, coho and spring, bear badges of black and white fin stripes or spots, vertical parr marks laterally, and orange flashes on operculae, fins, and jaws (Figures A2.4 - A2.5).

Territoriality also occurs during spawning, and is accompanied by dominance ranking. Spawning groups comprise a brightly colored dominant male (Figures A2.6 - A2.7) occupying a territory centered around a female, and one or more subordinates at the edge. Attacks are frequent, and threat displays consist of parallel swimming with dorsal fin raising. Subordinate males resemble females and have a broad lateral stripe. Each species has a species-specific coloration (Figures A2.6 - A2.7) based on surface color patterns extending into the mouth, which is gaped as a courtship signal but is normally slightly open and the inside visible in breathing behavior.

Schooling behavior first appears among downstream migrating juveniles (smolts) and consists of attraction and following the responses of whoever is in the lead. Movements are almost simultaneous, and schools intensify under attack from predators. Schooled fish break ranks, particularly as juveniles, to patrol in shallow water or to feed, or when feeding in open water and a school of prey fish is found.

Management considerations include the provision of suitable habitat for territoriality with protection from excessive competition as well as predation, the provision of suitable optimum waiting places with accessible food, and the elimination of conditioning to stimulii associated with artificial food supplies, especially when these mimic stimulii naturally acting as signs for predators.

VIII. Applications - fishing and management

Spawning is the stage of the life cycle most sensitive to disruption since the physical-chemical and social requirements are most constraining at this time. Management meets the concerns in various ways in which development of artificial spawning systems (streams or hatcheries), improvement of existing spawning grounds, protection from incompatible developments, and escapement of stock from the fishery are all critical. For the latter, the escapement must be from the mid-point of the run, to a maximum number and no more, otherwise repeated spawning on constrained beds by less productive late arrivals digs out the earlier laid better stock. Of course such mid-point stock may also be premium fish from the industry point of view. With mixed stocks entering a fishery, considerable complications arise in judging escapements, and there is risk of overfishing, even eliminating, smaller runs which enter the fishery simultaneously with the major stocks. Those small runs may be important to particular groups of fishermen.

Fishing technology relies on a knowledge of fish behavior, which is then exploited for human purposes. There are two main exploitable weaknesses shown by the fish in their sensitivities. They can be duped by a false lure which attracts them to a hook, and they can be

Pink fry

Length 1 inch (2-3 cm). Parr marks absent, dorsal surface dark green, ventral surface and sides silver, i.e. counter-shaded.

Chum fry

Length 1.5 inches (3-4 cm). Dorsal surface green, parr marks faint, ventral surface and sides silver, i.e. counter-shaded.

Coho fry

Early - length 1 inch (2-3 cm). Dorsal fin with conspicuous white and black leading edge, parr marks prominent, anal fin with extended tip, and white and black leading edge.

Late - length 3 inches (8 cm). Dorsal fin with black leading edge and white tip, parr marks prominent, anal fin with extended tip and white leading edge.

Sockeye fry

Length 1 inch (2-3 cm). Body spots small, parr marks small and irregular.

Chinook fry

Length 1.5-2.5 inches (4-7 cm). Dorsal fin with dark leading edge and white tip, parr marks prominent, adipose fin with dark trailing edge, anal and pelvic fins with white leading edges.

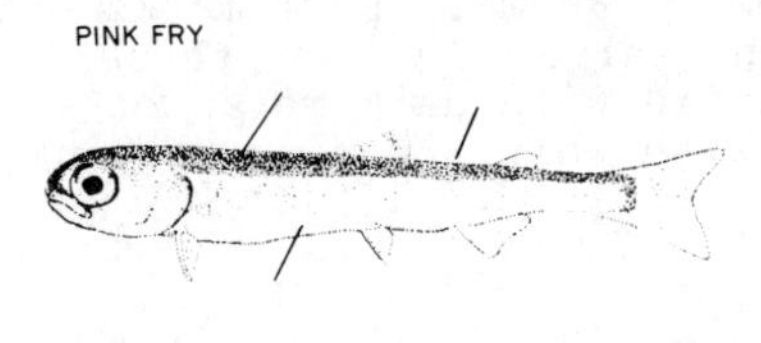

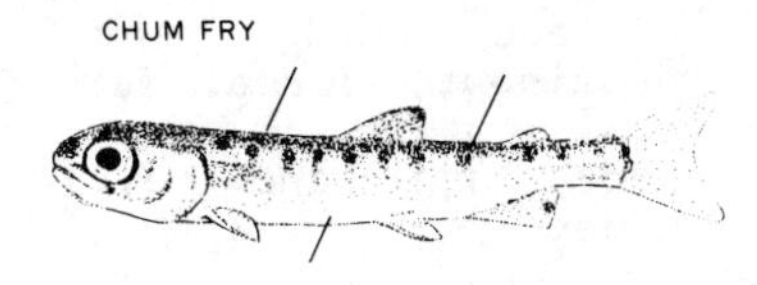

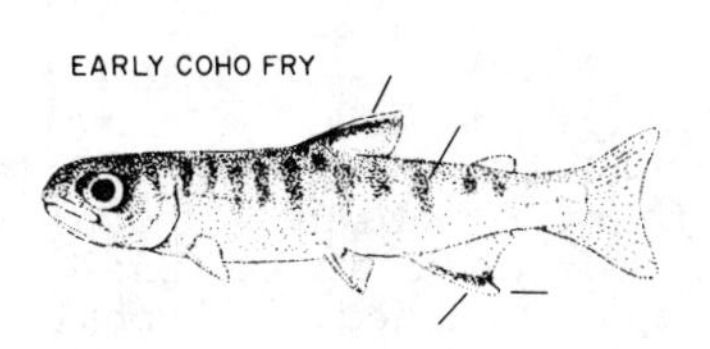

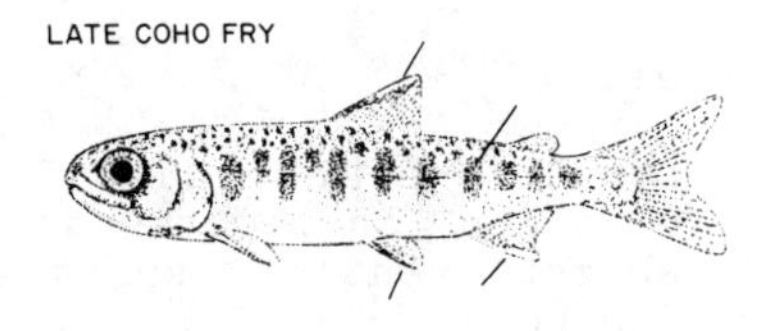

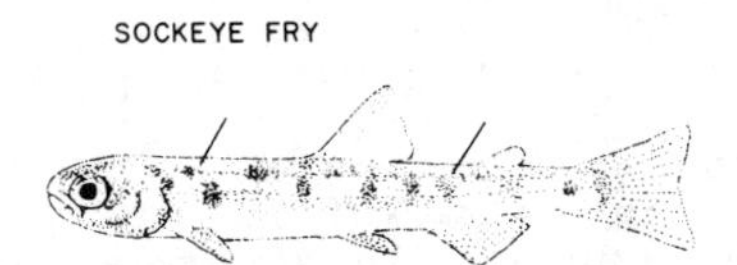

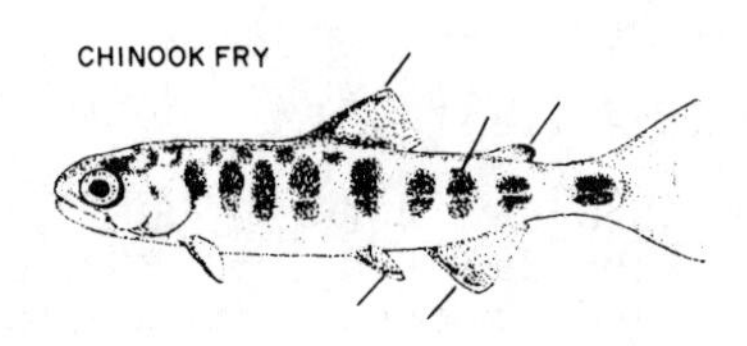

Figure A2.4. Morphology of Pacific salmon, indicating species-specific characters (fry).

Cutthroat fry

Length 1-3 inches (2-8 cm). Parr marks prominent, dorsal fin with white tip on three rays and spotted, adipose fin with dark marks on edge, anal and pelvic fins with white leading edges.

Rainbow fry

Length 1-3 inches (2-8 cm). Anterior dorsal surface with parr-like marks, parr marks prominent, dorsal fin with white tip on six rays and spotted, adipose fin with continuous dark edge, anal and pelvic fins with white leading edges.

Coho smolt

Length 4-6 inches (10-15 cm). Dorsal surface brown or green, dorsal fin with white tip and dark patch, parr marks faint or absent, caudal fin with dark patch at tip of each lobe, anal fin with white tip slightly extended, ventral surface silver, i.e. counter-shaded.

Sockeye smolt

Length 3-5 inches (8-13 cm). Dorsal surface brown or green, parr marks small, irregular and faint or absent, ventral surface silver, i.e. counter-shaded.

Cutthroat juvenile

Length 4-6 inches (10-15 cm). Body spots abundant, dorsal fin with white tip over three rays and spotted, caudal fin spotted, anal and pelvic fins with white tips, parr marks may be present.

Rainbow juvenile

Length 4-6 inches (10-15 cm). Body spots abundant, dorsal fin with white tip over 6 rays and spotted, red stripe on side of body, caudal fin spotted, anal and pelvic fins with white tips, parr marks may be present.

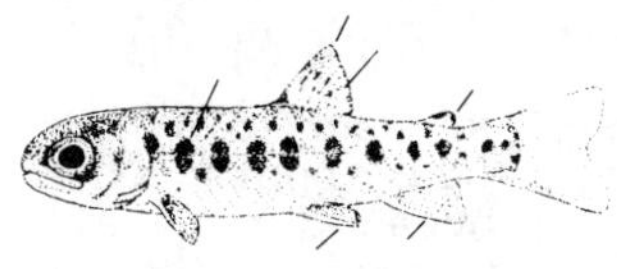

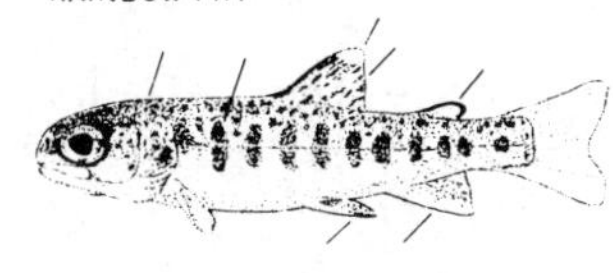

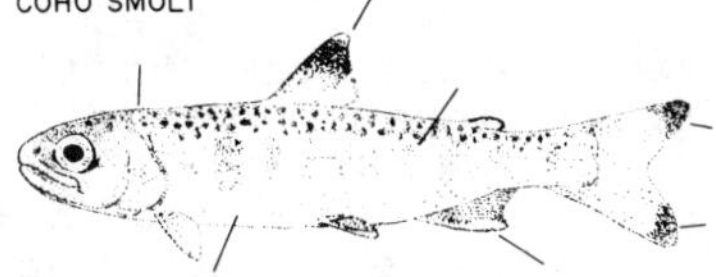

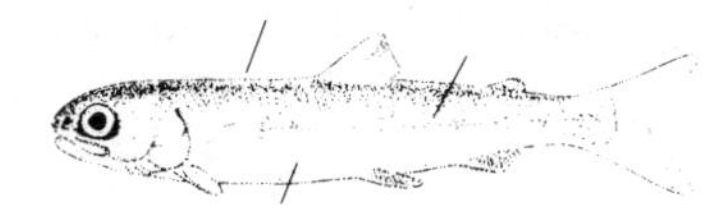

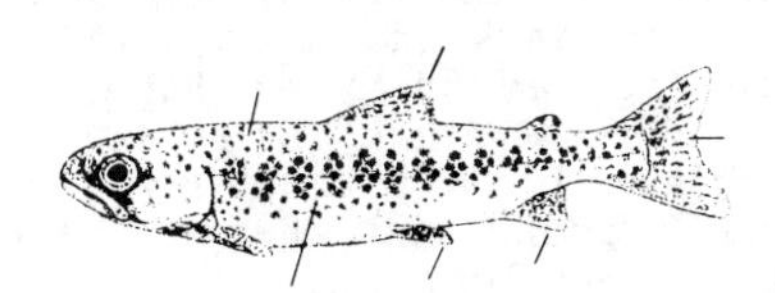

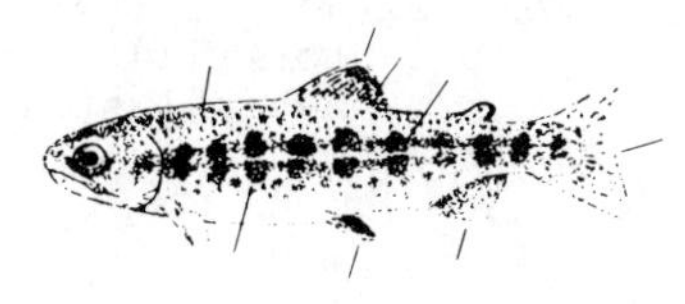

Figure A2.5. Morphology of Pacific salmon and regional trout, indicating species-specific characters (fry, smolts and juveniles).

Pink

Migrating - body spots present, side of body bears a broad, dark stripe, caudal fin spotted, ventral surface glaring white.

Male - anterior dorsal surface distended into a prominent hump, body spots present, side of body bears a broad dark stripe, caudal fin spotted, ventral surface glaring white.

Female (not illustrated) - resembles migrating form but darker.

Chum

Migrating - male and female spawning forms often assumed during migration.

Male - body spots absent, side of body bears vertical patterns of black, yellow and purple; prominent teeth.

Female - body spots absent, side of body bears broad, dark stripe.

Coho

Migrating - nostrils white, body spots present, dorsal surface dark (i.e. counter-shaded), caudal fin spotted on upper lobe only.

Male - nostrils white, body color very dark, side of body with broad red stripe, caudal fin spotted on upper lobe only.

Female (not illustrated) - resembles migrating form but overall color darker, broad red lateral stripe.

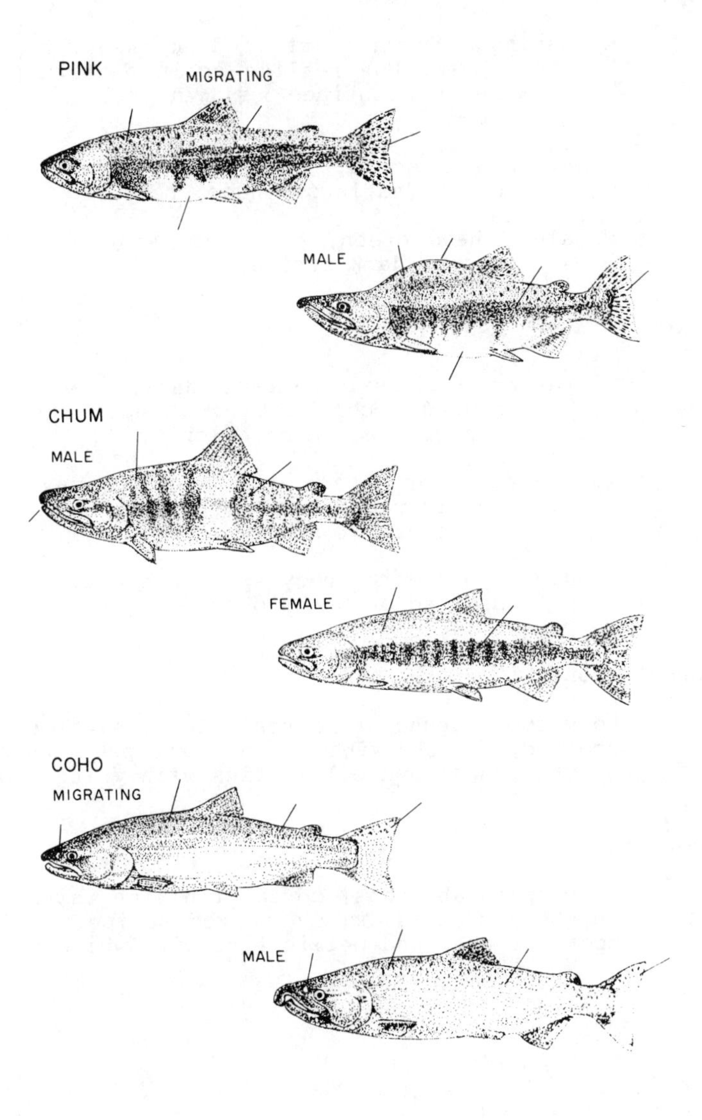

Figure A2.6. Morphology of Pacific salmon, indicating species-specific characters (Pink, Chum and Coho).

Sockeye

Migrating - dorsal surface dark green, body spots absent but netlike scale pattern present, ventral surface silver, i.e. counter-shaded.

Male - head green, anterior dorsal surface distended into a hump, body red.

Female - head green, body red, side of body with broad, dark stripe.

Chinook

Migrating - dorsal surface dark, i.e. counter-shaded, body spots few and large, caudal fin spotted on both lobes.

Male - anterior part of body may deepen considerably, body spots present, caudal fin spotted on both lobes.

Female - body dark, body spots present, caudal fin spotted on both lobes.

Cutthroat trout

Body spots abundant, dorsal fin with white tip and spotted, sides without red stripe, caudal fin spotted, anal and pelvic fins with white tips.

Rainbow trout

Body spots abundant, dorsal fin with white tip and spotted, side of body with red stripe, caudal fin spotted, anal and pelvic fins with white tips.

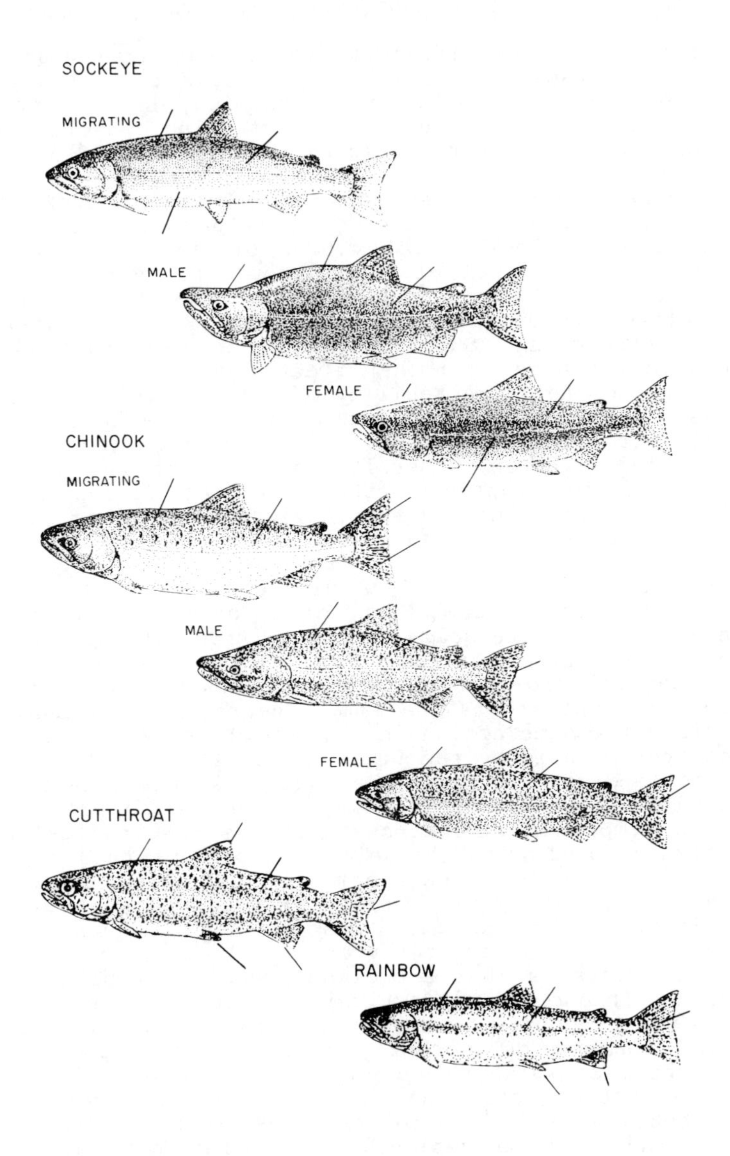

Figure A2.7. Morphology of Pacific salmon and regional trout, indicating species-specific characters (Sockeye and Chinook salmon, Cutthroat and Rainbow trout).

insufficiently responsive to a dangerous, human-made situation so that they do not escape in time.

Native Indians traditionally exploited these weaknesses with weirs, traps, and simple bone hooks on cedar lines. From these devices and western technology have sprung the existing seines, gill nets, the fishing wheel and an enormous variety of hook-and-lures (McKervill 1967).

The hook-and-lure type gears of commercial troller and sports fishermen have as an objective provoking the salmon to bite and get hooked. The lures simulate food, and are dependent on the level of the salmon's motivation to eat. They must not be used in such a way that they scare the fish, either through moving unnaturally or the smell of aversive chemicals. Surprisingly there seems to be little systematic research into the effectiveness of lures. This may be due to the intractability of the problem since salmon responses to food and lures can vary with the species, with time of year, with time of day, with satiation from previous feeding, and with many other less obvious variables. It has been shown, for instance, that identical lures used by skilled fishermen under strict instructions still give variable results between individuals (Argue 1970).

Purse seines and the now disused fishing traps have certain features in common, enough to group them both under the heading trap-type gears. Both rely on a social behavior pattern, that of schooling. In the context of a narrow river channel, "following" is a more descriptive term, but the behavior pattern is the same: the response of many fish to perform the same action in a group or in sequence. Purse seines are operated by the fisherman locating a large school of salmon and setting his net around it so that the fish are captured by closing the bottom of the net by the purse-strings. Implicit in this operation is that the seiner is not attracting the fish himself (but perhaps this may be a future technological development) but has located a naturally occurring and tight school (fish intensely responding one to another) and is able to surround it without provoking escape behavior. Implicit also in the use of purse seines is that a milling school will be easier to surround than one that is actively migrating. The shoreline fish traps in contrast relied on large moving schools and directed them by deflectors through a narrow door into a large trap from which they could be brailed out. The point in common between seine and trap is that they rely on the social behavior of salmon to stay close to one another or to follow, and must be designed and operated so as not to interfere with this social behavior pattern by provoking the opposed self-protective behavior of escape.

The remaining gear in use today, the gill net, can be characterized as an enmeshing type gear. Fish attempt to pass through the lightweight net but get caught by their gill covers. They push their heads through the meshes but cannot pass completely. As with trap type gear, gill nets must obviously not generate direct escape responses away from the net prior to enmeshment, but there are a number of unknowns about the extent to which other behaviors affect catchability. It is possible that schooling, or mass escape after some of the fish of a school have become enmeshed, could facilitate enmeshment of large numbers. It is believed that detection of gill net threads may be difficult for the salmon either visually or by touch, and that the initial response to slight restraints to swimming caused by the mesh is a thrust forward rather than a withdrawal. If so, the lighter, stronger and less visible the thread the better. Thus nylon nets are move effective than linen. On the other hand, schooling behavior may only be significant to the gill netter insofar as it concentrates fish along routes which can be predicted or detected by a good fisherman. Salmon behavior is so variable that perhaps we should consider the gill net as the general purpose gear which is not dependent on, although it will obviously benefit from, intensive activity. It should be noted, however, that a gill net fishery can be of such magnitude in a confined space (such as the mouth of the Skeena River) that it effectively acts as an enormous, ever changing, multi-walled floating trap (Todd 1969). In such cases it is dependent as with other trap type gears on intense activity of the fish, such as mass entry to the estuary from the sea.

The significant behavior patterns of the salmon for fishing can therefore be seen as feeding, schooling, activity level and escape responses. In general the last is to be minimized in the early stages of gear-fish interaction both by good design and operation. The other behaviors are important in that, when occurring intensely, the salmon become vulnerable to various gears. Perhaps the next technological developments will be to artificially induce intense activity in manners exemplified by the use of attractant lights at night and other forms of herding fish. Such developments are in hand and obviously need controlling as they progress.

The fishability of salmon is the composite result of their varying activities in juxtaposition with fishing gear. The fishery manager needs to appreciate the behavioral complexities of the situation and his appreciation must encompass the varying responses of different stocks in apparently similar situations.

APPENDIX 3. AUDIO-VISUAL PORTRAYALS.

Anyone interested in the behavior of animals needs to see their movements and hear their calls, either by watching living animals going about their daily business or by the next best -- watching a record of them on a cine-film, television documentary, videotape and/or listening to an audiotape. Good videotapes have come to be important teaching and learning tools for behaviorists.

This appendix presents a list of recent television series which show the behavior of various species in ways which are reasonably accurate and which have commentaries that inform reasonably objectively (Table A3.1). New television series will appear, as will new cine-films, and these should be viewed critically. There are some questions to ask when viewing a new film or television show. The ethogram checklist of Appendix 1 can be applied, but other questions are listed in Table A3.2. The intent of this checklist of questions is to allow a critical appraisal of whether the activities shown really reflect the behavior of the animals concerned.

There are also some working principles in using cine-films and videotapes to demonstrate behaviors to others. With modern playback equipment, videotapes can be edited during the presentation by using a playback counter. Thus in a long videotape, sequences can be selected to illustrate specific points. The user should prepare a set of notes with sequences identified by counter number and duration, so that he can edit as he plays back.

Many technically fine wildlife cine-photographers produce films and videotapes to which biologically inane commentaries are dubbed. These can be eliminated by turning down the volume control, and talking about the film yourself.

Always preview a rented cine-film or videotape and alert the class to interesting incidentals which otherwise might be missed. Short sequences of famous behaviorists, for instance, show viewers what the scientist was like.

Table A3.1. Videotape and cine-film series with behavioral emphasis.

1. "Life on Earth" series - David Attenborough.
2. "Nature" series - Donald Johanson (The Discovery of Animal Behavior).
3. "Survival" series - George Plimpton (Anglia TV).
4. "The Untamed World" series - Peter Backhaus TV Production.
5. "Wild Kingdom" - Mutual of Omaha.
6. National Geographic films.
7. "The World of Animal Behaviour" series - Jane Goodall-Baron van Lawick Films.
8. "The Nature of Things" series - David Suzuki.
9. Jacques Cousteau films: Cousteau Society, etc.
10. Nova: The Wild, Wild World of Animals - William Conrad.

I always ask students to list specific features of interest, such as the range of behaviors shown, and I leave sufficient light during the screening for note taking.

Table A3.2. Questions to be answered when viewing audio-visuals of animal behavior.

1. How many of the ethogram checkpoints from Appendix 1 can be answered?

2. How is the behavior described?
   Objectively (and interestingly)?
   Subjectively - in terms of emotions?
   Anthropomorphically - as though the animal were a thinking, emotional person?
   Anthropocentrically - with the animal's main reasons for existence to be available for exploitation by humans?
   Teleologically - with consequences of behavior claimed to be causes?

3. Is the process of observation and experiment described well?

4. How much irrelevant commentary or how many irrelevant sequences (even if beautiful or dramatic) are included?

5. Is the audio-visual more directed at the story of making the observations (or the audio-visual) rather than the content and significance of what was observed?

APPENDIX 4. OBSERVING ANIMAL BEHAVIOR.

For someone wanting to learn about animal behavior, it is essential to actually observe animals going about their daily business. The animals come alive, and no amount of written or spoken descriptions conveys all the many actions they are performing simultaneously as they balance the often conflicting demands of their ecosystem. Here is a set of observations which I have put together to make up practical assignments for the course that I teach in animal behavior. The assignments can be followed by any interested individual as well as by students under formal instruction.

Observations I. Social behavior: mallard ducks.

Mallards congregate at almost any park pond in northern North America. They are fearless of people under such protected conditions, and provided an observer does not feed them, they will perform whatever actions are appropriate to the time of year. They spread their breeding behavior almost throughout the twelve months, starting pair bonding in the fall, continuing through the winter, mating in the spring, and raising young in the summer. They should be observed at dawn and early morning prior to the arrival of the food distributing public, who then dominate the ducks' behavior and turn it primarily into begging. Some good sequence diagrams are available in Lorenz (1958).

Observations II. Social organization: tropical fish.

An aquarium of tropical fish provides an easy way to observe many different forms of social organization including territory, ranking, and schooling, mediated by threats or overt aggression. In a tank with many specimens time will be needed to recognize each as an individual fish, establish where they commonly place themselves, and discover their ranks within the groups which occur. Fish generally aggress one another by fin and operculum raising, and attack directed towards the tails of others. There are many descriptions of social behavior in tropical fish in books on their care and breeding.

Observations III. Invertebrate behavior: bees.

To some extent bees can be observed by standing close to hives and watching the bees come and go. They have a characteristic departure and arrival behavior which serves as an introduction to activities inside. To penetrate further a cooperative beekeeper is essential. He or she can set up a temporary transparent walled observation hive consisting of a comb, queen and workers. It is then possible to see the interior activities of dancing, communicating, and brood raising. By appropriate feeding, or observing when you know that food is being taken, the dances can be observed and interpreted. Descriptions are provided by Von Frisch (1953).

Observations IV. Comparative observations: ducks.

The pond of a public park in winter or a nearby bird refuge provides the observer with an opportunity to implement easily a comparative behavior study in which he or she looks for similar behaviors in different species and how they are manifested in each. If the species are closely related some of these behaviors will be homologous FAPs, such as courtship and agonistic displays. Descriptions are provided by Lorenz (1950).

Observations V. An easy ethogram: dogs.

The domestic dog provides an opportunity to easily construct an ethogram. In any suburban street where dogs are allowed to run free they set up territories, home ranges, friendship alliances and continuing hostilities. There is strong ranking and much social display, from the parade of a dominant to the urine marking of territories and ranges. The project provides an extended mini-field survey since the observer must be in position when the dogs are active, which is usually in the early morning in North American suburbs. Descriptions are given by Lorenz (1955) and Fuller and Fox (1969).

Observations VI. Learning and training: dog obedience classes or a dolphinarium.

An introduction into training and the way animals learn is provided by attending an obedience school with a dog. An extension to other animals is presented by a dolphinarium, where dolphins and other marine mammals are trained for public shows. The trainers at such dolphinaria have to use fairly obvious signals, and in some cases training and explanations of the training are open to the public.

Observations VII. Primates and zoos.

Zoos may provide an opportunity to observe several different primates and compare their behavior. Such observations are especially useful if one or another of the ape species is available. Commonly in zoos the species show strong male dominance, which limits the behavior to be seen. Observations at a zoo should be directed towards possible behavioral pathologies. Animals in bad physical condition, or those compulsively engaged in repetitive pacing and social aggression, indicate whether the observer will see very much of what is potentially a great range of behavior. Descriptions of zoo constraints on behavior are provided by Hediger (1950).

Observations VIII. Humans.

It can be a rewarding test of one's developing skill in observing animals' behaviors, signals, badges, and social interactions if one applies the techniques of unnoticed observation to groups of humans, thus developing a partial ethogram, or intensively describing some component of the total range of behavior. Ranking behavior within old or newly established groups at cafes or other meeting places is a favorite of those of my students who attempt this. Care has to be taken not to infringe on the rights of people to privacy, not the least because in some places they will adequately defend these rights (in a tavern) or there will be someone there to defend for them (in a restaurant).

Notes

1. The concept of biorhythm (Chapter 9) should provide an excellent medium for critically appraising those accounts of human behavior which are on the fringe of science. However, time is needed to establish whether one's patterns statistically fall into 23, 28 and 31 day

rhythms or not. It can be done, but it is time consuming.

2. It is desirable to observe the behavior of species which are locally important in any area. In the Pacific Northwest, Pacific salmon meet that requirement and it is easy to observe adults on spawning grounds in the fall, and juveniles in aquaria in the spring.

3. In general, I have found it difficult to get good observations of the animals which often attract student behaviorists. Deer and terrestrial mammals are not good, since even the best hunter-naturalists spend most of their time getting into position, and little actually observing. Some thought is needed in picking out good local species so that most of the effort is given to the actual observations, rather than the preparations. In general birds and fish are most useful. In addition to the species already presented, sea gulls can provide good opportunities. Garden animals, ranging from robins to spiders, are also practical.

INDEX